AF321686

Everyday Data Cultures

Everyday Data Cultures

Jean Burgess, Kath Albury,
Anthony McCosker and
Rowan Wilken

polity

First published in 2022 by Polity Press

Polity Press
65 Bridge Street
Cambridge CB2 1UR, UK

Polity Press
101 Station Landing
Suite 300
Medford, MA 02155, USA

ISBN-13: 978-1-5095-4755-5
ISBN-13: 978-1-5095-4756-2(pb)

A catalogue record for this book is available from the British Library.

Library of Congress Control Number: 2021951508

Typeset in 11.5 on 14pt Sabon
by Fakenham Prepress Solutions, Fakenham, Norfolk NR21 8NL
Printed and bound in Great Britain by CPI Group (UK) Ltd, Croydon

The publisher has used its best endeavours to ensure that the URLs for external websites referred to in this book are correct and active at the time of going to press. However, the publisher has no responsibility for the websites and can make no guarantee that a site will remain live or that the content is or will remain appropriate.

Every effort has been made to trace all copyright holders, but if any have been overlooked the publisher will be pleased to include any necessary credits in any subsequent reprint or edition.

For further information on Polity, visit our website:
politybooks.com

Contents

Acknowledgements

This project received support from the Australian Research Council Centre of Excellence for Automated Decision-Making and Society (ADM+S). The #spotify-wrapped Twitter data discussed in Chapters 3 and 5 was collected and processed with assistance from Betsy Alpert and the QUT Digital Observatory. The dating apps project discussed in Chapter 3 was supported by the Australian Research Council Linkage Project 'Safety, Risk and Wellbeing on Dating Apps' (LP160101687), in partnership with ACON Health and Family Planning NSW. Kath Albury and Anthony McCosker gratefully acknowledge their collaborators on that project, particularly Paul Byron, Teddy Cook, Christopher Dietzel, Tinonee Pym, Kane Race, Daniel Reeders, Doreen Salon, Son Vivienne and Jarrod Walshe. We would also like to thank the anonymous reviewers of the draft manuscript for their insightful, supportive and constructive feedback.

1

Introduction

A single parent is working from home as usual in Melbourne, Australia, connecting to the internet using one of the Google Wi-Fi points that are distributed strategically throughout her terrace house. She's prepared to trade off the privacy risks associated with the data these devices gather for the comfort she derives from their aesthetically pleasing sleek design, and for the convenience they provide compared to the complication and hassle of other home Wi-Fi set-ups, especially when renting. Meanwhile, two doors down, a man thumbs through search results and targeted ads on his phone as he researches the planned purchase of a new smart TV. He is struck by the low price of certain models, but feels uneasy, vaguely recalling reading somewhere that some models are only priced so competitively because the TV makers collect user data and sell it on to third parties (Gilbert, 2019).

In Madrid, a teenager opens Snapchat and scrolls through newly received snaps. Because his virtual presence on the Snap Map (his Actionmoji) is moving in a constant direction at speed, the Snapchat app correctly assumes that he is seated on a train. Beside him, his friend, who has been scrolling TikTok continuously, sees a video pop up in her 'For You Page'. It seems out of place compared to the videos the algorithm usually selects for her, and features a popular influencer humorously suggesting she take a break, have a drink of water and perhaps go for a walk. Barely pausing to roll her eyes at the clumsy intervention, she scrolls on to the next video in her feed (Burke, 2020). In a Brazilian megacity, a young woman is about to head out on her bicycle, and needs to work out the safest and most terrain-friendly route to take. Because she doesn't entirely trust app-generated directional recommendations, she selects her preferred route based on a combination of saved Strava and pre-downloaded Google Maps data and her own experiential knowledge of the city (Pink et al., 2019, p. 179).

Around the same time, a Japanese TV show reveals that the person behind social media star @azusagakuyuki, an attractive young woman who poses beside her motorcycle, is Soya, a fifty-year-old man. Soya created his highly successful female alter ego using AI-enabled face-editing apps – marketed as 'fun' apps, but licensed to users on terms that enable companies to collect large amounts of personally identifying data. When asked about it, Soya explains that, while he began by just playing around with the app, 'it happened to turn out to be fairly pretty' – and so @azusagakuyuki was born. Encouraged by the 'likes' he got after posting the results, he said, 'I got carried away gradually as I tried to make it cuter' (BBC News, 2021). Nearby in South Korea, Seo-Hyun, a young woman interested in fashion and beauty, gives careful consideration to the forms of 'zero party data' – 'personal data a consumer intentionally and proactively shares' with brands and marketers (Mitchell, 2019) – that she is prepared to share with clothing and cosmetics brands, tailoring this information in such a way as to maximise her chances of being rewarded with promotional give-aways of her favourite jeans and make-up.[1]

Outside a courthouse in Oakland, California, an activist taking part in a demonstration against police violence is confronted by a local officer, and so the protestor begins recording video of their interaction on his phone. The police officer retaliates by pulling his own phone out of his pocket. To the bemusement of the watching crowd, he opens Spotify and starts playing a track by mainstream pop artist Taylor Swift, assuming that YouTube's copyright enforcement algorithms (YouTube, 2021) will use audio data-matching to automatically detect and remove the protestor's video, preventing it from reaching a large audience (Cabanatuan, 2021; Schiffer and Robertson, 2021). Nevertheless, the video goes viral on Reddit and Twitter, provoking discussions and knowledge-sharing about audio editing techniques that can be used to work around the data logics of the major platforms' automated content-moderation techniques.[2]

These composite vignettes are a mix of factual, semi-fictionalised and fictionalised accounts, designed to give the reader (you) a way into the book's themes and ideas.[3] They have also fulfilled an important function for us as authors. In preparing this book, the creation of vignettes aided us in compiling and distilling what we have observed over the last decade or so of our own and colleagues' qualitative research in this area, and in isolating what we see as important about everyday data cultures. In this way, the above vignettes serve as 'presentation devices' (Ely et al., 1997, p. 74): they introduce and synthesise themes that are central to the book and to which we return, in different ways and to varying degrees, in later chapters. Not only do these vignettes reinforce how data is often both collected from and targeted to us as we go about our day-to-day lives (the datafication of everyday life), they also provide glimpses of how we form intimate relationships with and through data (everyday data intimacies), how we develop skills and capacities to do things with data (everyday data literacies), and how everyday data practices play out in communities and in public (everyday data publics).

The turn to data

At the most basic level, our opening vignettes together paint a preliminary picture of what is often referred to as the 'datafication' of everyday life. Our diverse daily activities – from connecting to the internet at home to managing our daily commute – are not only increasingly dependent on data; these activities and experiences are also routinely converted into digital data via our use of mobile devices, apps and the sensor networks embedded in our environments. And, increasingly, this data is used for automation and algorithmic decision-making processes, the results of which are fed back into our everyday lives.

As part of the first stage of this transition to datafication, services supporting a wide range of mundane activities and essential services (including media consumption, banking, shopping, transport, interpersonal communication and dating) have been steadily migrating to mobile apps and platforms with which we engage via data-driven identities and personal data profiles (as happens, for example, when we use Facebook to log in to a wide variety of other websites and apps). Meanwhile, an increasing number of data-collecting sensors and internet-connected objects are becoming embedded in personal devices, 'smart' homes, workplaces and public spaces. And digital media platforms integrate into their operations various forms of algorithmic curation and artificial intelligence, which feed on, learn from and act on this data, thereby shaping our access to information, our cultural experiences and our relationships.

Datafication represents a significant moment in the history of technology and society. For critical scholars, the concept of datafication expresses the capacity of 'commercial digital media [to] capture the details of activities that once eluded systematic forms of value extraction in order to turn them into information commodities' (Andrejevic, 2010, p. 90). Relatedly, datafication is thought to be transforming our identities as both consumers and citizens: the increased 'scope and sophistication' of data collection and processing has made datafication 'a cornerstone of contemporary forms of governance', enabling 'both corporate and state actors to profile, sort and categorize populations' (Hintz, Dencik and Wahl-Jorgensen, 2019, pp. 2–3). Mejias and Couldry (2019) put the situation in even more dramatic terms: for them, the term 'has quickly acquired an additional meaning: the wider transformation of human life so that its elements can be a continual source of data'.

But the datafication of everyday life, especially by state and corporate actors, is also a recent manifestation of longer-term trends. History is full of attempts to 'pin down' and fix as information the most difficult-to-grasp aspects of everyday life, precisely because everyday life was thought to be at the heart of what makes societies tick, how they change – and, hence, how their populations can be 'nudged' in one direction rather than another. For example, both Ben Highmore (2015) and John Storey (2014) devote entire chapters of their respective books on everyday life to the Mass-Observation project – the grandly ambitious attempt, undertaken by British (middle-class) researchers in the 1930s, to seek understandings of the lived experience of (working-class) everyday life by gathering first-hand information (from the 'masses') about it. The project generated large amounts of material, filling thousands of boxes with 'accounts of nightmares; meticulously detailed records of drinking habits in Bolton pubs (timed to the second with a stopwatch); pages and pages of diary records; [and] thoughts on margarine' (Highmore, 2015, p. 75).

And of course, audiences have been calculated (measured, segmented and targeted) by the media industries for as long as they have existed. In an earlier period of media studies focused on the challenges of studying television, John Hartley argued that broadcast television's audiences were 'invisible fictions' who had no existence outside the methods brought to bear on them by critics, policy-makers and industry actors. As Hartley points out, audiences need 'constant hailing and guidance' from the industry in order to see themselves *as* audiences – a system for 'imagining the unimaginable; for controlling the uncontrollable' (1987, p. 136). We see echoes of Hartley's ideas in the way that the large internet platforms have constructed, measured, targeted and addressed (hailed) us – these days, not as audiences but as 'users' – over the past forty years, continuing on the trajectory of inventing data and metrics (from likes to retweets) that can stand in for audience practices that might otherwise remain private and personal. In the post-broadcast era, where broadcast, print and internet media have converged, these processes of audience measurement and segmentation, and the metricisation of their attention, are intensified (Burgess and Baym, 2020; Livingstone, 2004, 2019).

In media studies, the turn to datafication is often discussed as part of the debate about a trajectory from mediation to mediatisation (for an excellent overview of the attendant debates, see Livingstone and Lunt, 2016). Since media are thought to not only represent but to help shape social realities, changes in media – understood both as communication technologies and as the symbolic representations transmitted by those technologies (Silverstone, 2005) – have always been connected to changes in society. The idea of a shift to 'mediatisation', though, is that the organising principles and values (the 'logics') of the media system begin to play a dominant role in society more generally (Couldry and Hepp, 2017). Building on Altheide and Snow's (1979) idea of 'media logic', José van Dijck and Thomas Poell (2013) discuss how characteristics of 'social media logic' – like data-driven popularity metrics – have come to influence far broader spheres of social and economic life. Most recently, some media theorists have proposed the idea of 'deep mediatisation' (Couldry and Hepp, 2017; Hepp, 2019), as a further intensification of these tendencies, and one particularly tied up with datafication and platforms (Andersen, 2018). Under the data logics of 'deep mediatisation', not only might a restaurant need to be visible on social media, the owners might even think about redesigning the space or plating menu items in ways that are specifically targeted to maximise engagement on Instagram, with the visual data processing and algorithmic logics of the platform in mind (as the restaurant web-hosting company *Owner*'s 'Ultimate Guide to Instagram for Restaurant Professionals' outlines in forensic detail).[4]

Importantly, it is possible for people to be affected by these processes of datafication and 'deep mediatisation' even when they don't have much access to or active engagement with digital technologies, meaning that we can be subject to these systems but have limited agency within them. Digital inclusion (and exclusion) remains an issue in almost every country (see, for example, Thomas et al., 2018), and in the context of datafication and automation, its implications for social inequality are only increasing. A lack of universally affordable, meaningful access to telecommunications data is one major source of inequality (Moyo and Munoriyarwa, 2021), intersecting with other aspects of digital inclusion or exclusion such as digital skills and confidence (Park and Humphry, 2019). Digital inclusion has particularly acute implications for people with disability, while at the same time their experiences afford 'a rich and indispensable site and "test bed" for how societies can confront technology for better futures' (Goggin, Ellis and Hawkins, 2019).

Our opening vignettes also hint at one of our principal interests in this book: the ways people live with, work around, or resist these data operations. A core argument here is that people aren't always only subject *to* datafication. Instead – in certain ways, at certain times and within certain constraints – we are active agents and, sometimes, disrupters and resisters *of* datafication as well. If nothing else, everyday experiences of technology are messy – even the 'smartest' of smart technologies never really work quite as seamlessly as either the marketing hype or the panicked media stories about them would have us believe. At a deeper level, everyday life is where the politics of digital transformation are worked through in practice, as people negotiate, wrangle, learn and struggle with or against data-intensive technologies, in the context of their own bodies, lives, communities and histories. We see these politics played out in different ways in the earlier vignettes: the Japanese biker's off-label use of a face-editing app; the young South Korean woman's canny manipulation of zero party data in the context of intensive ad targeting; and the cyclist's circumspection around the safe use of navigational apps in the city she knows so well.

'Everyday data cultures' is a cultural studies-based conceptual framework that can be used to explore how the broader digital transformations associated with datafication (or 'deep mediatisation') are playing out at ground level, and how the activities, thoughts and feelings of citizens, consumers and users of technologies play a part in these processes. In what follows, we discuss the many ways data is created, transformed and shared in and through people's daily activities. But in turn, people's everyday practices of and ideas about data influence and shape Big Tech's data logics, infrastructures and flows, adding friction and noise to their business and revenue models. This interplay between everyday cultures of use and business logics can be seen playing out in different ways in the earlier vignettes: in TikTok's health interventions and the user's eye-rolling resistance to them; and in police counter-tactics that seek, in the most mundane and off-hand manner, to weaponise automated platform responses to copyright infringement. These 'everyday data cultures', in all their variety, richness and political contradictions, are the focus of this book.

Critical data studies and beyond

In the current period of intensive activity and media coverage related to artificial intelligence (AI), algorithms and automation, there is a growing chorus of critics and scholars sounding the alarm. These voices articulate an increasing concern about the take-up and power of data-intensive, automated decision-making technologies, both in terms of their ubiquity and in terms of the new or intensified forms of inequality and injustice that can result from them (see, for example, Eubanks, 2018; Amoore, 2020).

Intervening in the previous period of hype around 'big data' in the 2010s, scholars critiqued the inflated claims for its analytical power and revolutionary potential (Puschmann and Burgess, 2014), the problems of bias in applications of AI, such as predictive policing (boyd and Crawford, 2012), and the way the apparent seamlessness of data operations obscures infrastructures of logistics and labour. This work was important in highlighting the gaps between public understandings of and industry hype around both the benefits and dangers of datafication.

Since that first wave, the field of critical data studies has emerged at the intersections of digital sociology, cultural studies and internet studies (Iliadis and Russo, 2016), and has made significant advances in theorising, and diagnosing the politics of, data and datafication (Cheney-Lippold, 2017). The collection, reuse and exploitation of personal data by both corporate and government organisations has provoked concerns about trust, privacy and surveillance, leading to calls for new data rights (Ruppert et al., 2017), improved data literacy (McCosker, 2017a; Fotopoulou, 2021) and shared principles for data ethics (Zwitter, 2014).

Data has a cultural politics, too. As Catherine D'Ignazio and Lauren F. Klein remind us in their book *Data Feminism* (2020), the politics of data lie not only in what it includes or leaves out, or even what it *does*, but in the very idea of data, with its connotations of objectivity, its evidentiary power, and the binary logics through which it is so often constructed and deployed. Prefiguring and accompanying concerns about the uneven benefits of AI and datafication, and the problems with fairness and bias that can result, important interventions have also been made that challenge the unthinking centring of whiteness in technoculture (Brock, 2020), the racial bias of much data-driven decision-making, and the real-world impacts of racist data discrimination (Benjamin, 2019; Noble, 2018). Such interventions have already made a noticeable impact on the public conversation in this area, and have provided the impetus for potentially significant change within the technology industry – see, for example, the acclaimed documentary film *Coded Bias* (Kantayya, 2020), which featured some of the most prominent critical social scientists and tech activists in the United States, including former Google employees.

This critical data studies conversation has recently and very noticeably gone mainstream, and in the process it has taken a distinctive turn towards what we call 'Big Critique', by which we mean writing marked by a sense of understandable urgency expressed in increasingly polemical terms, set up principally against the unprecedented power of the large technology companies – primarily US-based companies like Google, Facebook and Amazon. To a lesser extent in Anglophone scholarship, given its tendency to US-centricity, such concerns also include Chinese companies like Tencent and BytePlus.

Big Critique is often characterised by bold new concepts and ideas that describe large-scale, whole-of-society (or whole-of-planet) concerns. In *Atlas of AI*, for example, Kate Crawford (2021) addresses at planetary scale the profit-driven logics and damaging social and environmental impacts of AI and the data it runs on. Mark Andrejevic's *Automated Media* (2019a) argues that data-driven automation brings with it a deep paradigm shift, wherein logics of prediction rather than representation, and the frameless 'fantasy of total information capture' (Andrejevic, 2019b), threaten not only to further entrench surveillance but also to close down the space of possibility for political action.

Nick Couldry and Ulises A. Mejias's *The Costs of Connection* (2019) provides an expansive theory of datafication in terms of colonialism, and the threat to human autonomy and social life itself that, they argue, the 'coloniality of data relations' poses. We must remember here that – as Indigenous and Black scholars have long told us – information and data have always been tools of the colonial project. Indeed, Ian Pool calls 'the collection, use and misuse of data on indigenous people' colonialism's (and postcolonialism's) 'fellow traveller' (2016, p. 57) – always, though, met by resilience and joyful resistance (Carlson and Frazer, 2021; Brock, 2020). It is very important to keep this history in mind when Big Tech appears to make similar moves on the historically privileged among us.

A potentially serious problem with some of these forms of critique is that they risk echoing, mirroring or amplifying – rather than debunking – the dominant myths about the power of technology. This risk is heightened when critique becomes unmoored from specific, lived experience, and instead is discursively elevated to the same heights of global omniscience as Big Tech.

The clearest and most prominent example of Big Critique's tendency to mirror the rhetoric of Big Tech is Harvard business professor Shoshana Zuboff's epic – indeed, as anthropologist Anush Kapadia (2020, p. 33) has called it, 'operatic' – work *The Age of Surveillance Capitalism* (2019). In the book, Zuboff sounds the alarm about a new, aberrant form of capitalism that takes the excess data traces (or 'digital exhaust') left behind by our increasingly digitalised lives, converts them into predictive analytics, and then uses them for behavioural manipulation. Despite its overheated rhetoric, there is no doubt that works such as Zuboff's have highlighted increasingly urgent issues impacting society at both the individual and community levels and at planetary scale, and strong polemical language may aid in shifting the needle in more ethical directions (for an excellent 'review of the reviews' of *Surveillance Capitalism* that arrives at a similar conclusion, see Jansen and Pooley, 2021).

At the most polemical and populist end of the continuum, Big Critique is exemplified by the highly popular Netflix documentary film *The Social Dilemma*, released in mid-2020, and representing a particularly acute moment in a media environment awash in discourses of selfie narcissism, screen addiction, viral fake news and algorithmic radicalisation. The film centres on social media's algorithmic manipulation of user behaviour, and centres agency and responsibility almost entirely with the tech companies that provide some of the most popular social media platforms, primarily relying on tech insiders to diagnose the problems they claim (in hindsight) to have caused.

Despite the benefits of gathering attention for the cause, arguing from within the dominant framing (a pathological one, of addiction and behavioural manipulation), as both *Surveillance Capitalism* and *The Social Dilemma* do, risks becoming politically unproductive, because it discursively strips internet users – and all 'ordinary people' – of human agency. Not coincidentally, this framing also betrays the ongoing encroachment on media discourse of information-centred and behaviourist models of communication, so that we humans are cast as either the sinister agents or the unconscious subjects of behavioural tracking, targeting and manipulation – and that, of course, is exactly how Zuboff's 'surveillance capitalism' wants to see us. Therefore, in meeting hype with counter-hype (or, as Science, Technology and Society scholar Lee Vinsel [2021] calls it, 'criti-hype'), Big Critique can end up effectively reinforcing the claims that Big Tech makes about itself.

The pattern of media and industry hype and corresponding counter-hype around new technologies has a long history. The idea of a (permanent) technology 'revolution' is in turn articulated to deeply colonial ideas about the relationship between nationhood, progress and technology – particularly in the context of the United States, but with local resonances elsewhere. This cultural formation, which attributes awesome (or fearsome) power to AI, is the latest iteration of what David Nye (1996) called the American 'technological sublime', a framework later applied to the 1990s Silicon Bubble by Vincent Mosco (2005). Under this framework, the dominant narratives around data and AI as being all-powerful form part of the quasi-religious formation that is American technoculture (Nye, 1996; Carey and Quirk, 1970; Mosco, 2005; Brock, 2020), which represents itself as transcending lived experience, and therefore as somehow floating above the politics of race, gender and sexuality – which is to say, it is coded as white (Brock, 2020, p. 34). American technoculture leans so far into the digital sublime as to treat its male CEOs, from Steve Jobs to Elon Musk and Mark Zuckerberg, like demigods – an impression only intensified by their intensely self-promotional adventures, whether travelling to space in phallic rocket ships, or to the 'metaverse' in VR headsets.

When Big Critique heroically takes Big Tech on, fighting rhetorical fire with fire and revealing the toxicity of dominant technocultures, it plays into these religious tropes of magical technologies at the centre of moral battles between good and evil. As Luke Goode argues in discussing the mythos of AI, 'reflecting the polarities of the technological sublime, we see prophecies of doom vying with those of rapture' (2018, p. 200). The discursive pattern of hype and counter-hype (to which Big Critique contributes) ends up serving the interests of Big Tech, because it tends to invoke a sense of drama and urgency around an always-impending future tech revolution, and to paint a picture of future possibilities – whether utopian or dystopian – that assumes AI is able to do what it claims it can.

Meanwhile, individuals, communities and organisations are going about their lives and work amid constant technological change: grappling with, anxious about, joyfully resisting (Lu and Steele, 2019), or just not very interested in, the possibilities, risks and challenges of data and automation. In our choice to focus on these experiences and practices, we are closely aligned to the work of Helen Kennedy (2016, 2018), who has argued strongly in favour of 'inserting the everyday into data studies and data activism'. This approach requires far more attention being paid to the experiences, thoughts and feelings of non-expert citizens ('ordinary people', in the cultural studies sense) who are 'living with data and datafication' (Kennedy, 2018, p. 20). And, importantly, Kennedy and others (Lupton, 2018) deliver on this project by conducting empirical research that involves paying attention to the practices and listening to the voices (Burgess, 2006) of ordinary people.

But in the most prominent conversations at the present conjuncture of tech developments and social concerns – whether concerning AI, algorithms or, until recently, 'big data' – the actual practices, thoughts and feelings of audiences or users are too easily overlooked. Sonia Livingstone diagnoses the present moment as a 'heady climate', one in which 'cautious calls to gather evidence about people's lives are easily missed in the urgent rush to describe our coming predicament' (2019, p. 176). There is a history to this over-investment in critiques of technology production and corresponding under-investment in understanding how ordinary people might be affected by these developments. In the post-war era – a comparable period of intense rhetoric around the technological sublime in the Cold War United States – a major investment in mass communication and media effects research for propaganda purposes was coupled with a 'critical' tradition whose attention to media power came at the expense of an interest in audiences (Livingstone, 2019). But empirical work persistently showed that media audiences failed to fall into line: they simply did not behave like the easily manipulated 'mass' of either the propagandists' or the critics' imaginations.

In the present context, there are ample empirical accounts of social media users' reluctance, distrust and frustration (see, for example, Bucher, 2017; Light, 2014), and of the ways that pleasurable and connective uses of these technologies can serve progressive, strengthening, even radical ends (Carlson and Frazer, 2021; Lu and Steele, 2019). As Livingstone says, these and other empirically grounded accounts of resistive, tactical or refusing digital media users cast serious doubt on the assumption that platforms can do what they claim and many fear: the 'effective imposition of power' on audiences. Such work therefore gives 'encouragement to those' (such as Kennedy) 'who call for alternative approaches that respect audiences and publics' (2019, p. 180).

To extend these principles – of scepticism towards the idea of total media power, and of interest in the experiences of media subjects – to the era of datafication, we might consider targeted programmatic advertising. As everyone who has seen *The Social Dilemma* knows, advertising is the principal source of revenue for digital media platforms, and the principal site of social anxiety around 'surveillance capitalism' (Zuboff, 2019). All advertising relies on the buying and selling of audience attention – a commodity which is convertible to value through measurement – and therefore on datafication. Online advertising has become financialised and is now almost fully automated. But what if advertising doesn't actually work in the behaviourally manipulative ways upon which its market logics depend? That would leave the web's entire economy resting on shaky foundations, so that online advertising ends up looking more like a speculative bubble than an industry (Hwang, 2020). At the very least, as we know from media history, the reading practices of the audiences for these ads are likely to evade and exceed the advertisers' intentions (Livingstone, 2019).

Beyond textual practices like reading against the grain, audiences resist automated advertising in other ways. For instance, filtering technologies such as ad blockers serve as an example of friction introduced by user practices (Thomas, 2018), which sit alongside the use of proxies and other data-centred tactics (like password sharing) to circumvent the geoblocking of streaming services due to local licensing and regulatory restrictions on media content availability in local markets (Lobato, 2019). These sources of friction between platform logics and everyday cultures of use might end up, as Thomas (2018) points out, having the perverse outcome of disabling the open web as we know it and accelerating the trajectory towards platformisation (Helmond, 2015; Nieborg and Poell, 2018), with a further hardening of Facebook, Apple, Amazon, Tencent and Google et al.'s monopolistic tendencies. They might also lead to a new cycle of user refusal, circumvention and creative adaptation – or indifference.

In this book, then, we pull focus from the cycles of dramatic hype and counter-hype in order to closely observe people's lived experiences with data and data-driven machines, drawing on our own empirical work with people and their everyday digital practices, as well as close readings of online digital culture, and the work of many other colleagues. By paying careful attention to the particularity and specificity of everyday data practices, we aim to offer a discursive antidote to the dehumanising abstractions that characterise mainstream white technoculture (and upon which the digital technological sublime depends), as well as highlighting everyday practices that run counter to the dominant narrative. While directly addressing issues around surveillance and the inequalities arising from Big Tech's exploitation of data and algorithms, we also learn from the practical solutions cobbled together by suburban families, the ways that queer digital intimacies provide joyful and caring models of selfhood and relationships, and the ways that 'ordinary people', too, can perpetrate data harms.

We approach this task with a strong sense of hope. Our aim is to add a voice in counterpoint to the chorus of concerns gathered around Big Critique, and an alternative to dystopian views of how datafication is impacting on our lives closer to home. We want to show that, as well as being the zone of social life where datafication impacts particularly intensively on people, everyday life is an important source of mess and noise. In doing so, we hope to provide an alternative to the dominant narrative about the inexorable march of the sublime machines, and offer some conceptual tools that might mitigate against 'conceding the inevitability of finitude' (Andrejevic, 2019b) in scholarship, research and practice.

About this book

In focusing on everyday data cultures, we train the spotlight specifically on how people live with, through and as subjects of data as they go about their daily activities – at work, at play, at home and on the move. The people we are interested in are citizens and consumers, industry actors and government agents; they are babies, children, adults and seniors; their social identities, socioeconomic circumstances and geographical locations vary widely, as does the extent of their power and agency over the datafication of ordinary life and society. We're interested in their data practices and experiences not only as individuals, but also as participants in intimate relationships, communities and publics. Throughout, we give consideration to the intertwined relations between machine and human data practices. We seek to highlight and amplify everyday human agency, as well as explore its limits and uneven distribution, and consider how it is being transformed through the logics of data and automation.

In the chapters that follow, we further develop and elaborate the everyday data cultures framework, grounding it first in concepts of and approaches to studying everyday life drawn from cultural studies, and then through worked examples. Our examples are carefully chosen and diverse (within the limits of our frame of reference), to enable comparison and to illuminate the political tensions inherent in the intersection between datafication and diverse everyday lives.

In working our way through these examples, our approach mirrors Daniel Miller's call to 'de-fetishise' theory (that is, to negate its tendencies to 'abstraction, generalisation and de-contexualisation') through exemplification and comparison. In the journal article to which we refer here, Miller (2021) demonstrates this approach through his own collaborative, longitudinal work on the mobile phone and its cultures of everyday use. Similarly, we try throughout the book to present theory simultaneously as a scholarly conversation (through citations) and through real-world examples – so that, in Miller's terms, 'from the start it is clear what is meant by this theory as applied to some concrete instances, and also how one can envisage using that theory in other instances' (2021, p. 874). Indeed, we hope that by the time you reach the end of this book, many 'other instances' will have suggested themselves, to which our everyday data cultures framework can be applied, or against which it can be tested. In keeping with this approach, we use examples from popular digital culture, from domestic life, and from community interventions, which, we hope, will highlight the framework's adaptiveness and generativity.

Given all our talk about the importance of understanding everyday data cultures not in the abstract but in particular, we also need to situate ourselves and our collective speaking position. All four of us live in Australia, although one of us was born in the United States. We are white settlers living on colonised Aboriginal land, where the sovereignty of First Nations people was never ceded. Two of us grew up in country towns, but all of us have lived and worked in Australia's most populous East Coast cities for most of our lives. By virtue of education and occupation if not birth, we are all middle class; but before finding our way into academia we all worked in other jobs and professions. Some of us are parents, some are not; we vary in terms of gender and sexuality. All of us are aged somewhere between forty-five and fifty-five at the time of writing, placing us solidly in the demographic popularly referred to (to the extent we are noticed at all!) as Generation X.

We recognise both the limits and the specificities of these cultural contexts and social identities. This means we have reached first for examples that run close to our own experience with the internet and popular culture – many of our touchpoints are located in Australia, in the Anglosphere more broadly, and also in our own geographic region. But we also learn from and cite scholars who approach the issues covered in the book from a range of identity-based and international perspectives, including Black American and Indigenous (including Indigenous Australian) scholars, as well as work located in the Global South. Despite these efforts, we don't (and wouldn't want to) claim a universal perspective. We look forward to future dialogue with other scholars about where we could connect with their work, and what we've missed, as well as critical accounts of how well the everyday data culture framework applies in other contexts, and how it might need adjusting.

The structure of the book is as follows.

In Chapter 2, we begin by establishing the conceptual terrain of everyday data cultures. Outlining the key concepts from cultural studies that underpin our approach, we develop an understanding of this conceptual framework in two steps. First, we clarify what we mean by culture, how the everyday relates closely to it, and why it ('everyday life') matters. Second, we unpack our understanding of the 'ordinariness' of data. To set the scene and establish a new method of inquiry, we emphasise the twin dimensions of everyday data cultures – as the *cultivation* of data through everyday life, and as everyday *cultures of use*.

We then begin to explore the terrain of everyday data cultures from the inside out, beginning in Chapter 3 with everyday data intimacies, which are about the datafied self and its relationships. First, we discuss the ways that apps and platforms come to know their users as we generate, curate and share intimate data – and, in turn, how they invite us to use this data to know ourselves. Second, we examine the emerging sextech industry, the promised enhancement of intimacy through the datafication of pleasure, and its faith in data dashboards as tools for self-understanding and sexual fulfilment. Finally, we look at how data-intensive technologies are promoted as tools for 'careful surveillance' in homes, families and relationships, promising increased domestic safety and security at the same time as they facilitate coercive control.

In Chapter 4, we explore ways of knowing and learning in, through and about data cultures in everyday life, and the forms of agency associated with these everyday data literacies. As scholars building on Richard Hoggart's (1957) *The Uses of Literacy* have shown over many decades (see for example Pangrazio, 2016), literacies are at the coalface of individual and collective struggle – and so it is with data. Through cases spanning the various uses of self-tracking for personal well-being, workplace optimisation systems and the gaming of social media metrics, the chapter identifies the tactics, strategies and skilfulness through which people make do with, hack or even transform corporate and platform logics and operations through their personal and collective encounters with data.

In Chapter 5, Everyday Data Publics, we turn to collective responses to datafication and everyday data production as a reminder that data practices are not only personal or individual, but also collective and social. This final chapter deals with the changing ways we have come to know ourselves *as* publics, through multiple aggregates of data and data visualisations. We explore examples of the dynamic constitution of data publics (McCosker and Graham, 2018) that emerge, engage and dissipate on and across social media platforms, and that demonstrate their cultural and sociotechnical power – positive and negative, productive and destructive – in everyday life.

We close with a Conclusion that draws out lessons and insights from the cases and material covered throughout the book. One aim of this is to also draw on the 'resources of hope' (Williams, 1989) that are to be found in everyday life, highlighting attempts by civil society and other actors to chart a path through the political tangles of datafication and to enhance human agency and flourishing. Our closing contention is that this book, and the terrain it has covered, reveals to us how vital it now is to address the ordinary and everyday circumstances of data in building new opportunities for collective action, activism, scholarly inquiry and creative agency.

2
The Everyday Data Cultures Framework

In this chapter, we do four things: we establish the everyday data cultures framework and its basis in the cultural studies tradition; we detail what we mean by the 'ordinariness' of data; we show how an everyday data cultures framework assists us in understanding our day-to-day practices; and we demonstrate why this approach, and giving voice to the experiences of ordinary people, is both important and necessary.

Cultural studies and everyday life

In developing the conceptual framework of everyday data cultures, we are building on a long tradition within cultural studies. Over the second half of the twentieth century and into the twenty-first, British cultural studies and its divergent offshoots elsewhere (including in postcolonial settings such as Australia, where we learned it) fundamentally redefined 'culture', showed how it operates in and through everyday life, and placed it at the centre of studying societies and social change (Turner, 1990). Indeed, John Storey (2014) argues that everyday life has always been an object of study for cultural studies, and that it has latterly become its central object. Likewise, Langbauer argues that '"the everyday" is a foundational category in cultural studies', so much so that it is taken for granted and not often defined (Langbauer, 1992, p. 47). The first thing to say about everyday life is that, in centring it as our object of study, we refute the residual value judgements that cast everyday culture as inferior to formal artistic culture, or frame it pejoratively as the (mundane, banal, feminised) stuff left over from life's more worthy or significant activities. For us as scholars, and as people, 'culture is ordinary' (Williams, 1989), and everyday life is at the heart of culture.

A fundamental insight of the early tradition in British cultural studies, with its twin commitments to popular education and class struggle, was that while culture was often thought of as 'the best that has been thought and said' (Arnold, 1869, p. viii) – that is, elite, highbrow culture – it could also be understood as a society's 'whole way of life', with complex variations across social classes (Williams, 1961, p. 18). In this view, which we share, culture is practised and learned wherever there are people (and animals, plants and things), and it is through these practices that we come to be, understand and remember who we are (Storey, 2014, pp. 106–7). This is what Williams meant by the phrase 'culture is ordinary', drawn from his famous essay with that title (collected in 1989's *Resources of Hope*, but first published in 1958).[5] Culture, then, is both particular and shared – and we are all involved in making and living it. As Storey points out, however, 'this does not of course mean that we are all involved in it in the same way; meaning-making, like all other social activities, is always entangled in relations of power' (2014, p. 129). Further, as Williams wrote in *The Long Revolution*, culture is the site of historical change and struggle:

> Our whole way of life, from the shape of our communities to the organisation and content of education, and from the structure of the family to the status of art and entertainment, is being profoundly affected by the progress and interaction of democracy and industry, and by the extension of communications. This deeper cultural revolution is a large part of our most significant living experience, and is being interpreted and indeed fought out, in very complex ways, in the world of art and ideas. (Williams, 1975, p. 12)

Culture, then, is the complex and evolving system of meanings and values that lies at the heart of society. It is through culture that we collectively experience, make sense of and contest processes of historical change – including large-scale digital transformations such as datafication. These changes are negotiated and expressed through communication and media. Because of this, more obviously mediated and symbolic forms and practices are easier to see *as* culture (this is why we are used to thinking of film, music and television as cultural industries). But while some parts of cultural studies are primarily concerned with media texts and audience readings of them, it is not only through media or art that culture is made, transmitted and contested – culture is embodied, material and spatial as well, and everyday life is its main stage.

The everyday doesn't necessarily 'have a form of attention that is proper to it' (Clucas, 2000, p. 25), or any agreed-upon means of recording it, with key theorists diverging in the approaches they have taken:

Georges Perec was drawn to experiments in description, Roland Barthes to structuralist semiotic analyses of objects and cultural encounters; Henri Lefebvre advocated attention to 'moments', while Luce Giard undertook detailed qualitative research within households, and so on. But what is agreed upon is the need for attentiveness to everyday culture 'at the level of its living detail and actuality' (Highmore, 2015, p. 152) – that is, to everyday practices.

And yet the unruliness of the everyday presents particular epistemological and methodological challenges. How are we as cultural critics best able to register the everyday as a part of, or inhering in, 'manifold lived experience' without it dissolving into 'statistics, properties, data' when it becomes the object of study? (Sheringham, 2006, p. 360). One consequence of the elusiveness or indeterminacy of the everyday – what Sheringham (2006, p. 30) calls its 'fruitful ambivalence' – is that it never forms a stable and knowable totality. The everyday evades capture (Lefebvre, 2000, pp. 9–10; Blanchot, 1987; Martin, 2003, p. 2), including by bureaucracies, government and corporations, and resists various forms of reduction, including to statistics and data (Sheringham, 2006, pp. 17–18). K-means clustering algorithms (that partition a dataset into distinct subgroups where each data point belongs to only one group) and nearest neighbour algorithms (that assign items within a dataset to a cluster by determining what other items are similar to it) can be effective tools when calculating mobile media users' past, present and future movements in time and space; and services like Google and Foursquare City Guide routinely use these techniques in their recommender systems (Barreneche, 2015). However, while they are used in order to build geodemographic profiles of users and their socio-spatial patterns and behaviours, these data portraits are not effective in capturing the full richness and messiness of the events, rituals and everyday encounters that attend these movements and interactions. And yet it is precisely this elusiveness and familiarity – the indeterminacy of the everyday – that points to its significance and suggests that these everyday routines and actions add up to more than the sum of their parts: 'the everyday is the accumulation of "small things" that constitute a more expansive but hard to register "big thing"' (Highmore, 2010, p. 1). This dynamic between detail and big picture, and the significance of both, is nicely captured by Henri Lefebvre when he writes: 'the everyday is ... the most universal and the most unique condition, the most social and the most individuated, the most obvious and the best hidden' (1987, p. 9).

Georges Perec, an indefatigable examiner of the everyday, never stopped advocating for the need to reach beyond the 'big event, the untoward, the extraordinary' (Perec, 1999a, p. 209) – a shift captured in his aphoristic imperative call: 'not the exotic any more, but the endotic' (p. 210), the opposite of the exotic, the ordinary and that which is inside the ordinary (the 'infra-ordinary'). He asks:

> What's really going on, what we're experiencing, the rest, all the rest, where is it? How should we take account of, question, describe what happens every day and recurs every day: the banal, the quotidian, the obvious, the common, the ordinary, the infra-ordinary, the background noise, the habitual? (Perec, 1999a, pp. 209–10)

Such questioning is not intended as a paean to the mutability and apparent evanescence of the everyday.

Rather, what is being advocated for here is a renewed and focused attention on the rich variety of everyday practices.

Despite the rather abstract and generalised way some critics talk about 'the everyday', which can have the effect of erasing identities, bodies and histories from the picture, we take care to recognise the ways that particular everyday lives (or, perhaps more appropriately, lifeworlds) are grounded in (post)colonial histories and specific material conditions (Sandywell, 2004). These contexts reinvigorate our understandings of 'culture as ordinary' (Shome, 2019) and re-emphasise the rich pluralities of everyday data cultures. Perec's is a call that also implies the requirement to speak from lived experience, rather than observing the behaviour of the exotic other from a great height – heeding the lessons of the Mass-Observation studies examined by Storey (2014, pp. 40–53), in which the lives of the lower classes were fixed with the colonial gaze of the middle class.

If it can be done without imposing the disciplining logics of the colonial gaze from 'above', then paying attention to everyday practice can reveal a great deal about the particularities of individuals and their everyday lives (our utensils and tools, the way we spend our time, our rhythms). At the same time, it can also reveal a lot about collective practices, or what Highmore (2015, p. 147) terms 'collective intimacies', and why it is that we might seek solace in 'highly selective communities on social networks who might share a huge amount of similar likes and dislikes, of cultural references, of values' (p. 146). It tells us why it is that we might mourn the loss or decline of particular platforms, like Tumblr, that have fostered such collective cultural intimacies (Tiidenberg, Hendry and Abidin, 2021).

However, as de Certeau, Giard and Mayol observed, everyday practice 'is relative to the power relations that structure the social field as well as the field of knowledge' (1998, p. 254). This is to say that the possibilities presented by the everyday occur within certain forms of constraint and subject to power differentials. There are manifold examples of such constraint involving digital technologies, including the introduction in Australia of phone bans in some public schools, prohibiting the carrying and use of mobile phones during school hours, and felt pressures to convert 'unproductive' or 'wasted' commutes into work-based 'productive travel time' (Lyons and Urry, 2005). It also encompasses workplace bans on the use of social media or on posting certain forms of content to accounts that may identify one's employer, as well as pressure to submit to various forms of workplace biometric measurement and surveillance – such as the requirement that Amazon delivery drivers consent to the installation of AI-powered Netradyne surveillance cameras in their vans or risk losing their jobs (Vincent, 2021b). Henri Lefebvre might have characterised these sorts of restrictions as embedded within 'linear' rhythms, which he associates with everyday human activity that is associated with 'the monotony of actions' and of 'imposed structures' (2004, p. 8). But, as Lefebvre makes clear, 'linear' rhythms are also subject to various forms of 'interference' (p. 15) or disruption, as everyday life is just as commonly characterised by 'eurhythmia' (regular rhythms), 'arrhythmia' (irregular rhythms) and 'polyrhythmia' (multiple, simultaneous rhythms) (p. 16).

Thus, for de Certeau and collaborators, everyday practice 'patiently and tenaciously restores a space for play, an interval of freedom, a resistance to what is imposed (from a model, a system, or an order)', and speaks to the 'inventiveness of users' (de Certeau, Giard and Mayol, 1998, p. 254). Writing in the mid-1990s, de Certeau, Giard and Mayol illustrate this point via the forms of what we might now call 'vernacular creativity' (Burgess, 2006) made possible by VCR players and stereos, where 'people record fragments of program, produce montages, and thus become producers of their own little "cultural industry," compilers and managers of a private library of visual and sound archives' (de Certeau, Giard and Mayol, 1998, p. 254). In this book, we consider contemporary forms of the same spirit, such as TikTokers' gaming of algorithms to achieve viral hit songs. This element also speaks to the way that the everyday can be regarded as a 'reservoir of dissident political activity' (Sheringham, 2006, p. 19). Similarly, in later chapters we explore various forms of activist resistance to dominant data cultures and associated forms of data capture and information profiling. It is worth noting, too, that resistance to dominant data cultures is not always overtly activist in orientation. It can also be manifest in more mundane acts, including digital disconnection or data withdrawal. For example, many Australians have refused to sign up to the government-run online medical database My Health Record, citing concern over the sharing of personal health data between government agencies without consent, or that data might be shared in ways that could prove discriminatory (Newman et al., 2020, p. 21).

Everyday life is rich with examples of deep signification whose meaning and politics might be harder to tease out: the culture of one code of football as compared to another, for instance; or everything involved in an 'ordinary' weeknight dinner of 'meat and three veg' in Anglo Australia as opposed to France, where each meal is said to be buttressed by bread and wine (de Certeau, Giard and Mayol, 1998, p. 85). But it is because such mundane practices *are* meaningful and encode values that we can speak of the culture of everyday life. The culture of everyday life is enacted, not in the abstract, but through *practices, material things* (Highmore, 2010; Perec, 1991; Barthes, 1993; Miller, 2008), *spaces* (Lefebvre, 1991; Perec, 1999b; Massey, 2005; Storey, 2014, p. 134) and *places* (Malpas, 1999; Perec, 2010) – and these things and spaces are increasingly internet-connected and datafied. We can see all these ingredients at play in the following example of a truck driver in Guangdong province, China, whose truck doubles as his home, and who has just finished a gruelling thirty-six-hour, 2,500 km journey from the central province of Hubei to the Pearl River Delta. After a few days' rest, he will set out on the return journey, in a trip that will be repeated four times a month. He worries that he will soon no longer be able to afford to make these journeys – the fees he pays via the electronic toll collection system, which bills him for using the road at various increments along his journey, continue to climb steeply because of the debt-financed construction of these roads (He, 2020).

While the everyday, along with analogous concepts like the banal and the mundane, is often used to mean the boring, uninteresting or unremarkable ('the rest, all the rest' [Perec, 1999a, p. 209]), it makes just as much sense to think of everyday life as rich with material practices and thick with feelings. It can be a source of comfort and connection – even enchantment (Highmore, 2010) – as well as drama, fear and oppression; and, even in its spatial form as 'the home', it is not a private enclave, but rather is 'powerfully shaped by broader social currents' (Shun-Hing, 2010, p. 714). As decades of feminist scholarship have highlighted, everyday life is a site of political struggle, as well as a powerful site of resistance to oppression (Langbauer, 1992; Felski, 2000; Shun-Hing, 2010). In addition to bringing many new digital conveniences and 'smart' devices into the home, the domestic intensification of data and AI may serve to exacerbate persistent gendered inequalities around domestic labour (Strengers and Kennedy, 2020, pp. 46–7). It also provides tools to facilitate the abusive practices of coercive control that are characteristic of domestic violence (Bailey et al., 2021), and that are every bit as embedded, mundane and routine as the practice of making coffee in the morning (Dragiewicz et al., 2018; Strengers and Kennedy, 2020, pp. 198–202).

It is the unspoken, unremarked and yet complex nature of the everyday, with its dense web of material practices (from cooking the weeknight dinner to gaming on the couch or setting up the home security system), that makes it so important as a site of datafication and its politics. The increased entanglement between the practice of everyday life and digital technologies is deeply significant for understanding the digital society in which we live, precisely because datafication is itself becoming so ubiquitous and so ordinary. If we lose sight of the everyday, we miss the greater part of datafication's social dynamics and material effects. But more than that – we sail straight past the very place where people feel these processes most deeply, and where they have the most capacity to act upon them, albeit in constrained and ethically complicated ways.

Data *as* everyday culture

This book is partly a response to industry hype and public concern about AI technologies, and an attempt to bring those debates down to earth. In alignment with the rhetoric of the digital technological sublime discussed in the previous chapter, AI technologies are often represented in industry and popular media discourse as operating autonomously, even existing in a post-material or 'virtual' domain (like the outdated notion of 'cyberspace'). But this image couldn't be further from the truth. In addition to requiring huge amounts of computational power and energy, and being entangled in complex ways with the people and institutions who develop and use them, these technologies have throughout their histories to date been fundamentally – and greedily – reliant on data, much of it labelled by hand and originating in the very material and messy contexts of everyday life. As Helen Kennedy (2016) has argued, data is ordinary.

For example, many of the core features of the major digital platforms – like Facebook's targeted advertising, Instagram's personalised curation of user feeds, Spotify's Discover Weekly playlist or YouTube's 'up next' feature – rely on algorithmic recommender systems, which operate over the sea of content and curate it for, or target it to, each individual user. These recommender systems in turn use various forms of machine learning, whose models are trained on existing datasets (images, music, text) and then usually combined with dynamic data signals generated through everyday platform use – our own user profiles, patterns of use and connections to other users.

In fact, ImageNet, one of the largest and most significant datasets used for machine vision (e.g. to recognise and categorise objects and human faces within images), was directly built on photographs scraped from all over the internet – many of which were uploaded by ordinary users as part of their everyday cultural practices. To search for the images, the scraper used keywords from WordNet, a structured database of English words which, as it turned out, came pre-programmed with some seriously skewed cultural values, in turn priming ImageNet to generate racist and sexist associations (Crawford and Paglen, 2019). The creation of ImageNet also relied heavily on crowd-sourcing the painstaking manual work of labelling the datasets, relying on the labellers' vernacular knowledges of the everyday scenes and objects in the images, as well as the application of cultural values and judgements (for an excellent genealogical overview, see Denton et al., 2021).

As these examples show, data is *cultural* – that is, data is the result of, and an agent in, processes of meaning-making and struggles over values. This holds true even when data is used to represent apparently objective phenomena like the weather. We might think of something like daily minimum and maximum temperatures at a given location as a 'natural' property of that environment – and of course, to be clear, heat exists whether or not humans are around to observe it. But in order to gather data about the temperature in that place, we need, at a minimum, a structure for the data (that is, human-generated categories such as date, time and temperature that we want to create records for), a tool and/or a technique for measuring (generating data about) the phenomenon we have come to understand as 'temperature', a format for tabulating the data (plain text and numbers encoded in a particular way, arranged into columns and rows, say), and some way to record (or store) the data. None of these utterly banal conventions are natural, of course; instead, each of them has had many alternatives, and they have been experimented with, developed, argued about and eventually embedded in technologies, practices and institutions over centuries.

In the process, diverse cultural ideas about heat were both influenced by and played a role in shaping the technologies used to measure it. As John P. McCaskey has argued in his conceptual etymology of temperature, 'how people thought about thermometers influenced how they thought about temperature; how they thought about temperature influenced how they thought about heat; and so on' (2020, p. 442). For example, McCaskey argues that for Aristotle and into the Western Renaissance, hot and cold were not measurable quantities but 'primary and independent qualities' – more like the discrete but complementary taste sensations of sweet, salty, sour and bitter than opposing ends of a continuum (p. 406). By contrast, in our post-thermometer world, hot and cold are zones on a linear scale that has come to be organised around the boiling and freezing points of distilled water. Technologies of datafication, and the logics of the data they collect, are not just about objective facts: they influence and are shaped by culture. This applies not just to weather, but to our understandings of sexuality and gender – as we discuss in Chapter 3.

Data-reliant technologies (including measuring systems and scales) are the imperfect outcomes of human struggles, professional competition and controversies, often involving awkward experiments amid the messy realities of the natural world (Edwards, 2010). Some of these struggles remain unsettled to this day, as anyone who has tried to convert weather predictions from Celsius to Fahrenheit, or vice versa, knows all too well. And the messy realities include, among other things, the obstinate refusal of water to settle on a fixed boiling point that could serve as a reference point for the temperature scale (for a comprehensive account of thermometry's fascinating history, see Chang, 2004). If even water won't behave in a predictable fashion, spare a thought for those attempting to extract consistent, reliable data from humans.

Turning human activities and attributes into actionable data has a similarly long history. The extraction of personally identifying data for commercial, governance and other purposes long pre-dates the arrival of digital platforms and Big Tech companies. Here we offer two brief historical examples, nine centuries apart, to illustrate the point. The first example concerns the Domesday Book. Recent research on the 1085 Domesday survey – an eleventh-century form of census describing in detail the landholdings and resources of England – suggests that William the Conqueror ordered it for two reasons: first, to gather 'more information about England's resources and to generate increased income from them' in order to pay for a strengthened army brought in to ward off external threats to his rule; and second, in order to record, legitimise, reinforce and stabilise 'the transfer of these resources from the English to their conquerors' (Dalton, 2021, p. 56). The Domesday survey has been described as 'an extraordinary, carefully choreographed assertion of royal might, designed to make the king's authority manifest in every honour, shire, hundred, manor and household in the kingdom' (Stephen Baxter quoted in Dalton, 2021, p. 35). From today's viewpoint, it is also a remarkable undertaking in that it 'embodied virtually all the ritual and documentary elements of contemporary land conveyance customs' (Baxter quoted in Dalton, 2021, p. 35), prefiguring census methods, cadastral surveying and other modern technologies of population governance.

The second example relates to the twentieth-century interest in the 'scientific management' of the workplace. In his study of modern office systems and information management, Craig Robertson develops the concept of 'granular certainty', which he coins in order to capture 'the drive to break more and more of life and its everyday routines into discrete, observable, and manageable parts' (2021, p. 3). Robertson links this concept to Frederick Winslow Taylor and his workplace-related time studies. Taylorism (as it became known) can be understood as a form of datafication. As Tim Cresswell argues, at the heart of the Taylorist project was 'the transformation of learned habit' – that is, embodied, habitualised, everyday work practice – 'into a rigorous and scientifically coded abstraction' consisting of data about human motion (2006, p. 86). Cresswell describes how Taylorist time studies 'sought to provide a system whereby the worker's subjectivity could be controlled' (p. 92) through the capture of data on everyday bodily movements in order to achieve greater worker efficiencies through the reorganisation (and mechanisation) of the body in time and space. These efforts were extended through Frank and Lillian Gilbreth's experiments with photographic micromotion studies (pp. 98–106). These were used in order to make the 'bodies of workers in factory and home intelligible in new ways' (p. 121) as data subjects, and to discipline, standardise and regulate work- and home-based time and motion. The 'datafication' work of Taylor and the Gilbreths serves as a precursor to more recent human factors research, a field that is also preoccupied with granular certainty and data-driven research on micro-movements and efficiency gains (Wilken, 2017). The earlier time-motion studies also presage the restrictive workplace methods employed in contemporary

Amazon warehouses, which are explored further in Chapter 4.

Robertson's concept of granular certainty has applicability beyond twentieth-century time-motion studies and workplace systems. Granular certainty can be seen to underpin more contemporary forms of datafication. The concept speaks, for instance, to contemporary attempts at creating 'statistical bodies' (Abend and Fuchs, 2016) through self-quantification (Jethani, 2021; Neff and Nafus, 2016; Lupton, 2016), an endeavour with its own lengthy history (Humphreys, 2018). The search for granular certainty also motivates the forms of data mining undertaken by tech companies (such as Facebook, Google and Foursquare) that seek to generate nuanced geodemographic profiles and implement predictive analytics based on the traces of our digital passage (Smith, 2019; Barreneche and Wilken, 2015; Barreneche, 2012).

Returning to contemporary concerns about the datafication of identity and relationships via platforms like Facebook, it is important to remember that even at the most banal level the forms of personal data these platforms collect and exploit are not natural, pre-existing properties of individual persons, but rather are relational – historically constructed and shaped by the platforms themselves. Additionally, the demographic data associated with a single individual may vary according to who is collecting the data and why, with the competing interests of platforms, users and advertisers frustrating the dream of 'granular certainty'. For example, on Facebook it is possible for users to enter a range of gender identity descriptors, matching the social identities they want to use for themselves; but, as Rena Bivens found in her (2017) research on the 'back end' of the platform, only binary gender is available to advertisers, which makes sense given how most consumer-oriented marketing still segments around binary gender, but makes less sense in relation to the cultural transformations occurring around gender in society more broadly. This example shows that 'gender' isn't a coherent piece of information that can be extracted from us; rather, in digital society our gender is co-constructed individually, culturally and technologically, and is subject to multiple forms of conflict and change. Associative, vectoral logics are increasingly at work in these processes, especially where neural networks are involved; in other places, fixed identity categories (as for legal identity) apply. Datafication is nuancing and complicating identity as much as it is fixing it with increased granularity.

While much of the anxiety about datafication concerns personally identifiable information, data is also an increasingly prevalent and material part of everyday life more generally. Whether ticking off shopping lists as we wander around the supermarket, trying (and failing) to sync the family or household Google calendars, stockpiling downloaded movies and wedding photos, or wrangling spreadsheets in the office, we live with and through data, and it can seem to pile up around us and become unruly (for discussion, see Kennedy et al., 2020, pp. 256–68).

It is this unruliness of everyday data that leads to considerations of 'what it feels like to live with data' (Pink, Lanzeni and Horst, 2018, p. 2), of the uncertainties and anxieties attending the 'messiness' of (digital) data, as well as of associated concern over possible data loss. One way that such uncertainty manifests is in relation to the untidiness of data and the lack of clarity around how it is being, or ought to be, handled. In a project two of us (McCosker and Albury) are involved in, a food security and recycling charity worker reflects on the messiness of day-to-day data-collection processes:

> I think, working with volunteers, in my experience it can be challenging if you don't have really good systems to collect the data in the first place. Sometimes down at my building I hear 'we've run out of this form' and the next minute it's on a scrap of paper with a tally. And it's like, 'where's that gone?' And if you really need to have your data to communicate impact, and it's buried in five stacks of paper with a scratched-on tally, it's not great. I think there's probably a lot of small organisations or older organisations that don't have technology awareness, that don't know how to collect their data well or are skilled in doing that. (Workshop participant, 5 May 2021)

Another way that uncertainty manifests is as hesitancy around learning new forms of data collection, and resistance to the idea of the inevitable march of technological progress and of seamless data capture. In a project another of us (Wilken) is involved in, which addresses technological obduracy in healthcare contexts, a participant tells of how a CEO at a regional hospital set about investigating why staff within a central triage point had not adopted the secure messaging system they had been instructed to use. They discovered the following:

> Apparently the triage administration people had made a unilateral decision that they weren't going to deal with this system because it was too complicated, and that they had then two systems in place at the same time. They just wanted to go to the fax, and that was all they wanted to deal with, and so [the new secure messaging system] just gradually faded away. (Participant interview, 4 March 2021)

Paying close attention to everyday practices – and the desires, uncertainties and anxieties that attend these – can reveal a great deal about the scrappiness of our lived data experiences, and of our engagements with the textures and traces of everyday data. Such attention can also assist in developing strategies and tactics to empower humans as we learn to live with data.

In addition, it is important to remember that data is also shared and mobile – it travels independently of us (between devices and platforms) and with us (on portable devices). This is as true of the smartphone as a 'container' (Sofia, 2000) of social media content, photographs and text messages, as it is of the portable hard drive or humble USB stick. With respect to the latter, we drop them in jacket pockets and bags and take them with us, we gift them, we keep them in drawers or stack them carefully in shoeboxes for possible future use, we pass them between colleagues and friends where work and leisure-related content is exchanged (Kennedy and Wilken, 2021), and we preload them with televisual content and swap, share and rent them within informal media economies (Pertierra, 2017, p. 15).

Data, in these examples, is far from abstract: it is *stuff*. Storing, sorting, using and sharing this unruly stuff is also part of the practice of everyday life. These mobile, material practices that blur the boundaries between private and public data throw into stark relief the distance between corporate and governmental fantasies of 'granular certainty' and the reality of everyday life.

Conclusion

In this chapter, we have set out the conceptual framework that anchors this book. We have explored the intellectual sources and tributaries that feed it, and we have elaborated on its component parts. We have also sought to draw out what is at stake in examinations of everyday data cultures and what this reveals about the rich messiness of our mediatised and datafied day-to-day lives.

Drawing on the cultural studies traditions discussed earlier, we have defined everyday life as the social space in which culture – our collective system of meanings and values – is embedded, contested and transformed through social and material practice. Everyday life may be ordinary for those who live it, but it is always particular – our lives are varied, specific and diverse. For our purposes, this helps us see how everyday data cultures are situated in specific sociotechnical and sociocultural contexts – they are grounded in the cultures of use of mobile apps and a range of mundane smart devices, and in the 'platform vernaculars' (Gibbs et al., 2015) of various social media services. Everyday data cultures take on distinctive characteristics in urban or rural settings, and in the lives of people of different ages, races, abilities and gender or sexual identities. Throughout this book, our examples touch on these diverse experiences of and with data.

Datafication is an ordinary part of everyday life – increasingly common, but not bland, banal or necessarily even benign, just as everyday life is not purely bucolic and benign. Many people's everyday lives are spent dealing with a constant undercurrent of systemic racism or domestic violence. Data-intensive technologies play a significant role in these circumstances too, as we discuss in Chapter 3 in relation to research into 'smart home' surveillance, and sexual racism in dating app cultures.

Consequently, we are concerned with data's everyday *cultures of use* – that is, how data is encountered, experienced, exploited and resisted by users in the practice of everyday life, and how vernacular norms and practices for data ethics and safety are being managed, tested and contested within and among user communities. In this way, everyday data cultures can be a zone of encounter between historical trajectories, social groups and their values – as we have discussed in previous work in relation to the data cultures of dating apps (Albury et al., 2017, p. 2), and which we give particular attention to in our discussion of everyday data literacies (Chapter 4) and everyday data publics (Chapter 5).

Data's cultures of use both complicate and introduce optimism and hope into the critical data studies picture, as we discuss in the Conclusion. Since people (as consumers, users, citizens and subjects) are actively or incidentally involved in 'domesticating' and contesting the meanings of datafication, we can equally be involved in reshaping those meanings towards civic/progressive ends. Part of our responsibility here, then, is attending to the ethical and political dimensions of everyday data cultures, including harmful data practices perpetrated by 'ordinary people', as well as the more complex forms of connection, care and control that can be animated or intensified through the uses of data.

Building on this material in the chapters that follow, we offer a portrait of everyday data cultures in all their quotidian complexity and diversity. The journey begins with the datafied self and its relationships (everyday data intimacies), through to vernacular knowledge practices (everyday data literacies), collective action and activism (everyday data publics), and, finally, to the possibility of more hopeful data futures.

3

Everyday Data Intimacies

In vernacular expression, the term 'intimacy' is often used as polite euphemism for sexual activity. But it is also associated with the emotional interactions between friends and family members and professional care-givers, involving verbal and non-verbal communication as well as physical touch. In both these contexts, intimacy can be understood to encompass experiences ranging from trust, care and playfulness to disclosure, vulnerability and the shared, mundane routines of household life enacted behind closed doors.

Looking across these definitions and understandings, there is a tension implicit in the notion of intimacy. On the one hand, intimacy suggests reserve or inwardness: the private, the domestic and the personal. And yet, to borrow Lauren Berlant's words, once shared with others, 'the inwardness of the intimate is met by a corresponding publicness' (1998, p. 281). Thus, on the other hand, 'intimacy also involves an aspiration for a narrative about something shared, a story about both oneself and others' (p. 281) – or what is elsewhere referred to as 'collective intimacy' (Highmore, 2010, p. 146).

Some forms of collective intimacy are enacted through our affiliations with public affinity groups (as explored in Chapter 5). Others are expressed through friendships, sexual and romantic connections, kinship networks or membership of family groups (whether biological or not), and shared domestic spaces. Offices, factories and other workplaces can also be sites of intimacy, where co-workers form friendships, engage in interpersonal conflicts and build political alliances while navigating employer demands based on gener-alised performance goals and metrics. As Highmore notes, collective intimacies involve both practices of self-presentation and expressions of care, affection or connection (2010, p. 147).

In the contemporary context, all these forms of collective intimacy are practised online. A growing body of scholarship focuses on 'digital intimacies', exploring the ways digital technologies are reshaping our under-standings of our innermost selves and identities, and offering new opportunities to communicate, connect and form relationships across apps, platforms and devices (Dobson et al., 2018). The reason we are devoting an entire chapter of this book to intimacy, though, is that within digital cultures these practices and relationships are both generated through, and grounded in, data. After all, intimacy is about knowledge (of the inner self, and of the other), and in contemporary societies, ever-increasing data is the key to that knowledge.

These developments, of course, have their histories. Colin Koopman's Foucauldian genealogy of personal information, *How We Became Our Data* (2019), traces the development and societal embedding of mundane formats like birth certificates and personality tests to highlight the dependence of modern selfhood on the data that identifies and, increasingly, profiles us. It is here that, critics of datafication tell us, data has most recently been intruding, going way beyond names, birth dates and social security numbers to incorporate ever more implicit, embodied and formerly 'hidden' signals picked up by sensors and incorporated into the data flows of surveillance technologies. But data categories, relations, logics and flows also structure the ways we encounter and get to know others – we only have to think about how, to 'match' us with potential partners, dating apps use algorithms operating over data we have contributed about ourselves or that has been imported from other platforms.

In this chapter, we examine a range of forms of intimate data cultures. Each brings a different focus to, and engages with specific understandings of, data intimacies. We begin with a consideration of data selves – that is, how we come to know ourselves *as* ourselves through our own personal media use and interactions with intimate data, and how platforms and other tech companies seek to capitalise on and extend our engagements with intimate data. As part of this focus on data selves, we consider Spotify and its interest in algorithmically promoting 'networked intimacy' (Chambers, 2017). We ask: how do apps and platforms like Spotify and TikTok come to know us – and shape how we know ourselves – as we generate, curate and share data?

We then give extended attention to sextech start-ups and their attempts at promoting 'data-driven intimacy' (Flore and Pienaar, 2020). We offer a critical reading of how the 'utopian, optimism-sustaining versions of intimacy' (Berlant, 1998, p. 282) promoted by sextech often serve to reinforce normative understandings of gender and sexual intimacy and pleasure, and how they reproduce pre-digital modes of datafying bodies. From there, we examine how dating apps seek to cultivate intimacy, particularly through the requirement for users to share significant personal data, and how this plays out differently depending on the apps involved. We close the chapter with an exploration of what Hjorth et al. (2018) refer to as 'mundane intimacy' – that is, how data intimacies play out in the context of everyday interpersonal relationships within households, and specifically when household members wish to keep tabs on one another through forms of 'careful surveillance'.

Intimate algorithms

Algorithmically curated and personalised digital media platforms are embedded in the practice of everyday life – and, arguably, music listening is the most personal and ubiquitous of the cultural forms that we now experience in this way, extending into the most intimate spaces. The Spotify streaming audio platform promises us the ability to 'soundtrack our lives' (Spotify, 2021). Accelerating the trend (which began in earnest in the 2000s) away from recorded music consumption being structured around albums, and instead towards apparently infinite streams of music (Wikström, 2019), streaming privileges the arrangement of music into professionally curated playlists organised according to its own emergent logics of genre, mood and situation; all of which in turn mediate between user tastes and practices, marketing categories and the logics of the record industry (Prey, 2020). Through its interface design and features, Spotify (and other platforms like it) simultaneously offers the appearance of free choice from a limitless library of available music (and podcasts), and helps order this sea of choice through intensively personalised content curation – a process of 'algorithmic individuation' (Prey, 2018).

Users help cultivate Spotify's algorithmic logics through ordinary practices that also feed data signals back to the platform – such as listening to (or skipping) tracks, adding (or deleting) playlists to or from their in-app 'libraries', following (or unfollowing) artists and liking or saving (or 'unliking') tracks and albums (Eriksson et al., 2019). The implication of the tagline 'soundtrack your life' is that the platform's modes of curation can enable us to access exactly the right music at exactly the right time and place, no matter the setting, occasion, activity or mood. As Maria Eriksson and Anna Johansson (2017) explore in their detailed study of Spotify's situation- and mood-specific playlists, the platform addresses its users both intimately and with a 'prescribed temporality', associating different musical styles and affective states with mundane activities marking different points in a (normatively described) day – inviting users to 'wake up and smell the coffee' with cheerful folk; 'keep calm and focus' on study or work during the day with chillhop; and finally curl up on the sofa with some 'late night jazz'. Spotify offers us both convenience of access and a kind of algorithmic magic: the promise is one of a potentially perfectible intimate correspondence between our music and our interior lives, as long as we actively and carefully cultivate our presence (and our data profile) on the platform. In so doing, it plays a role in *programming* our most mundane and intimate everyday activities.

Spotify has for several years been integrated with Facebook's 'federated' user identity service (Bucher, 2021, pp. 101–2), enabling ease of access via the familiar 'log in with Facebook' button, and thereby connecting

Spotify user profiles to Facebook's data ecosystem. While much of this data connectivity takes place on the back end of both Spotify and Facebook, it also enables one of Spotify's few social affordances: on the desktop version of the app, users can see what their friends are listening to, generating a kind of ambient intimacy (this feature is not available on mobile, however). Less obviously, associative relationships between users and between artists, based on shared musical characteristics, combine to influence our experience of Spotify, including the music suggested to us; and these connections in turn shape the cultural logics of the platforms, such as its playlists and the proto-genres (from 'sad indie' to 'red dirt country' and 'late night jazz') they invoke.

Our listening habits also influence how we are treated by the data markets that connect through Spotify (especially if we are using the ad-supported version of the platform), since our listening and user profile data are subject to a complex web of data sharing among platform companies (in this case, Facebook), advertisers, the recorded music industry and the satellite ecology of audience data intermediaries (Eriksson et al., 2019, pp. 167–72; van der Vlist and Helmond, 2021; Bucher, 2021, pp. 101–2). This will be more obvious to users of the 'free' version of Spotify, who receive ads that are the product of automated ad targeting based on data mixing (or 'alchemy' as Adrian Athique [2018] calls it), taking inputs from user demographics, location and playlist names and contents to serve up more-or-less targeted ads (although, perhaps assuming that Spotify itself is selecting ads for them, users often remark on the bluntness or flat-out wrongness of this targeting [Eriksson et al., 2019, pp. 167–72]). When we 'soundtrack our lives', then, even in the quietest moments of solitude, cocooned by noise-cancelling headphones, and even in the most remote locations, we are immersed not only in sound but in social data relations.

Spotify also offers users the ability to curate their own listening experiences via the creation of custom playlists, which appear in search results alongside the professionally created playlists that are heavily emphasised by the user interface as the principal mode of experiencing music on the platform. The process of curating custom playlists involves searching for songs that the user wants to add, selecting songs from the ranked search results, and perhaps adopting the platform's further suggestions about what songs might belong in the emerging playlist. These everyday practices of creative curation, tied as they are to our lived experiences, feelings and moods, both help feed and are shaped by Spotify's algorithmic recommender systems. Marika Lüders (2021) has shown how Spotify users exploit the playlist creation feature as a way to work against the grain, pursuing a 'will to archive' their music and as a way to connect their music more deeply to their own life histories, despite the platform's constant algorithmic nudges to discover endless new artists and tracks.

Spotify users exploit the playlist feature in highly creative ways: assembling mixtapes of favourites dripping with nostalgia, capturing the zeitgeist with TikTok trend-themed musical moodboards, or putting together 'bootleg' soundtrack albums of favourite Netflix series, and in so doing porting the platform vernaculars and cultural logics across platforms, cross-fertilising cultural data operations in the process. This data work of playlist creation is also a form of identity work (Hagen, 2015), in terms of the self and in terms of the communities that self is affiliated with. In their study of LGBTQ-themed user-created playlists, Dhaenens and Burgess discovered that this feature, which appears to offer 'curatorial freedom and an endless library', was used to build queer archives and make political statements. But in tension with this was the algorithmic tendency of the platform to drag the user's curatorial choices inexorably back to a fairly predictable 'canon of Western-centric LGBTQ music culture that revolves around dancefloor fillers, mainstream female popstars and LGBTQ anthems' (2019, p. 1208).

Increasingly, Spotify also explicitly invites users to accept deep personalisation as a mode of self-knowledge as much as music discovery. Building on the highly popular Discover Weekly playlists, launched in 2015, the platform serves up a weekly new releases playlist (Release Radar), as well as ad hoc personalised playlists like Time Capsule (an assumedly nostalgic playlist related to the user's age and listening history). At the end of each year, the platform drops Wrapped, an interactive data dashboard summarising the user's listening history over the past twelve months, with a focus on listening time, artists and genres. As part of this annual digital media event, the platform also nudges users (by way of an auto-post button) to share their personal dashboards and analytics via social media (Weiss, 2018).

Spotify Wrapped was supplemented for the first time in 2021 with a mid-year data profile (also accompanied by interactive dashboards and prompts to post to social media) called Only You (Spotify, 2021), which is explicitly focused on presenting the user with a (flattering) data portrait of their unique listening patterns and taste profile. In 2021, the platform also quietly launched 'Blend', which connects each user's individual Spotify data selfie with that of their friends. This latter development is notable because it runs against the platform's general tendency *against* social interaction on the platform (apart from the ability to engage in forms of 'ambient intimacy' [Reichelt, 2007] by following Facebook friends and observing their listening habits in real time).

So, through its data operations and the way they are presented back to us, Spotify represents itself as knowing us – in all our perfect uniqueness – effortlessly, playfully, intimately and benevolently. To use Spotify intensively in daily life, then, is to be constantly nudged to know yourself more deeply in all your uniqueness through data analysis of your everyday media use, which in turn is used to further personalise your experience of the platform: it becomes a resource to connect with, or differentiate the self from, others.

But each year's batch of #spotifywrapped tweets demonstrates the active, knowing and ambivalent relationship users have to Spotify's data-driven story about who they are as music consumers. The tweets always contain markers of the feelings that users have about their music and the artists who make it. These range from simple shout-outs to 'top' artists (as represented by each user's Spotify Wrapped data dashboards) to emotional expressions of gratitude for the music that got them through the year (and this was especially poignant in 2020). But there are also self-deprecating remarks about what each person's annual listening patterns say about them, and, in 2020, reflections on the strange temporality of a year in pandemic time – and how a person's embodied memory of the role that music played in their life during that time might contrast strangely with the supposed 'objective truth' of the listening data according to Spotify.

The #spotifywrapped tweets also express political critiques of the relationship between Spotify data, algorithmic calculations and the music industry (noting the relationship between an obsession with audience engagement metrics and a decline of cultural value, for example). Most intriguing in the context of this book are the comments on the glitchy relationships between actual everyday practices and the algorithm's logics – from tweets wryly noting Spotify's focus on that one nursery rhyme that is the only thing stopping a toddler from screaming every time they get into the car, to the white noise track that is needed to stave off insomnia, and the teens' house party tastes 'ruining' their parents' carefully curated indie rock audio profiles. In deliberately 'making over' the suggested format of the #spotifywrapped tweets that the platform nudges them to post, Spotify users highlight the gaps between their algorithmic profiles and the messy complexity of everyday life. In doing so via the public hashtag, though, they engage in practices of collective intimacy and take part in processes of collective learning and debate about algorithmic culture.

Similarly, TikTok's 'for you page' (referred to in the platform vernacular as the FYP) – the default home page on the app – is intensively personalised and algorithmically curated. The recommender system that drives the FYP is based on a discrete product developed by Chinese parent company ByteDance and deployed in other apps and services prior to TikTok (Kaye et al., 2021). In June 2021, the system was spun out into a separate, licensable suite of AI-driven personalisation and recommendation tools managed by the new BytePlus division, with customers including online travel, fashion and retail businesses (Bradshaw, 2021). As well as being TikTok's principal engine of economic value creation, it is the cultural heart of the platform. Far from being a hidden infrastructure layer, the FYP is often referred to by users (including both creators and audiences) in reified terms as 'the algorithm' – an autonomous agent that has power over users' experience of the platform, but that can be 'trained' through their usage of it (Sung, 2020). In vernacular discourse, both on the platform itself and in ancillary media accounts, it is not only reified but also granted sentience: 'the algorithm' is a discrete actor that can make or break creator careers, while serving up personalised content to audiences, profiling them in ways that range from spookily accurate to insultingly off-base.

Some user reflections on the ways the TikTok algorithm seems to 'know' them are wryly amusing, reflexively acknowledging the idea of a proliferation of algorithmic taste-, interest- and desire-based identities, and performing and building knowledge about algorithmic data operations in the process. User reactions to the algorithmic mirror TikTok's FYP holds up range from playful faux-embarrassment to discomfort with the data-driven self that is presented to us, often using affective terms commonly associated with technology (Paasonen, 2018), such as calling the algorithm 'creepy' or 'scarily accurate' (see, for example, Kaitlyn Tiffany's [2021] piece in *The Atlantic*, entitled 'I'm scared of the person TikTok thinks I am'). Such reactions have even extended as far as the idea that TikTok has (accurately) diagnosed some users with ADHD – an idea so common that there are TikTok trends (the platform vernacular for memes) devoted to it (Bosely, 2021).

These vernacular discourses call on familiar tropes of uncanny artificial intelligence, invoking myths of technologies that are definitely not human, and yet have invaded and can manipulate the world of human experience. This uncanny, transcendent quality is a feature of the digital sublime, and is a common theme in popular and critical discussions of algorithms, as scholars such as Nick Seaver (2018) and Blake Hallinan and Ted Striphas (2016) have argued. As Seaver has also explained, in reality there is very little that is artificial about algorithmic decisions, including the decision about what song, image or video should appear next in a user's feed. Rather, there are, in the words of a former Spotify executive, 'humans all the way down' (Seaver, 2018) – every algorithmic data operation and automated decision also involves human practices (including not only those of the company and its engineers, but also those of its users) and is embedded in social relationships.

These algorithmic imaginaries of sublime machine intelligence feed the interests of the platforms, which do, after all, want to be seen as magical in certain ways – perfectly built, all-knowing and seamlessly usable. But this aura creates tensions, because they also want to be seen as friendly, comforting and familiar – embedded in, and a cosy companion to, our everyday lives. This is why the use of data to aid intimate self-knowledge and social connection – to help users see the platform as something that can soundtrack their unique, quirky and yet ordinary lives – is a vital part of initiatives like Spotify Wrapped, and why they are essential to the platform cultures of digital media.

Data is the material through which digital and social media platforms mediate between 'the self and the world' (Burgess, 2015), and platform data operations like algorithmic recommender systems thicken this web of relations. But as we have seen with responses to Spotify Wrapped, there remain plenty of gaps and loose threads – enough for platform users to selectively resist, critically reflect on, or take pleasure in the ways that platforms try to know us, and invite us to know ourselves better, through data.

Sextech's data intimacies

As Brady Robards and colleagues (2020) have demonstrated, digital platforms and devices have supported and promoted a proliferation of everyday gendered practices of intimate datafication and self-tracking. We will discuss self-tracking as a practice of data literacy in depth in Chapter 4. Here, we focus on it only tangentially, as one element within the emerging field of sextech, which has become a site where intimate data cultures are both cultivated and contested. While sextech is not yet mainstream (Bluetooth sexual devices are expensive and relatively niche), it represents a space in which everyday understandings of intimacy and data privacy are publicly contested, as both tech users and popular commentators push back against reports of data brokerage and data leaks (Sundén, 2020).

Sextech serves as a catch-all term, applied to dating apps, sexual entertainment platforms and services (such as OnlyFans and camming sites),[6] networked sex toys and AI-powered sex robots. It overlaps with femtech, a term that is applied both to sex toys and apps and platforms supporting sexual and reproductive health and well-being (for example fertility trackers). While datafied sex toys (such as butt plugs, vibrators and penile masturbation sleeves) fall under both the femtech and sextech umbrellas, they are framed quite differently in these intersecting domains. Sextech devices pitched towards men are frequently (though not exclusively) marketed as enhancements for pleasure and opportunities to playfully experiment with self-tracking in a leisure context.

In contrast, sextech products and services marketed to women (particularly those that are aligned with femtech) are more likely to be presented as tools for self-care and well-being. In these contexts, data collection and self-tracking is less an opportunity for recreation and more akin to a health regime. Indeed, the FemTech Collective, a networking association for entrepreneurs in the field, describes femtech as 'a category of software, diagnostics, products and services that include … fertility solutions, period-tracking apps, pregnancy and nursing care, women's sexual wellness, and reproductive system health care'.[7] As Lupton observes, while sextech gamifies men's sexuality, it tends to medicalise women's experiences of sex and reproduction, while simultaneously offering the promise of objective 'data-driven' understandings of sexuality (2015, p. 336).

But whether presented as digital sex toys or sexual well-being devices, networked toys offer users an opportunity to generate, organise and 'play' with data. For example, the Lioness vibrator brands itself as 'the world's smartest vibrator for better orgasms', offering 'biofeedback' by tracking vaginal contractions during arousal and orgasm. The Lioness website frames the 'rabbit' style vibrator as a quasi-therapeutic device that elicits what might be termed a quantified sexual self. Users are invited to 'experiment with different factors (Kegels, alcohol, moods, health, toys, sensations and more) and see how your pleasure changes'.[8]

With a sensor that records vaginal contractions and visualises them via an associated app and dashboard, the Lioness is an exemplar of the sextech/femtech crossover. Users are explicitly invited to contribute their data to sexuality researchers to build collective knowledge of sexual function, via a research portal that invites users to 'do it for science', by donating their recorded data, and participating in surveys and other studies. The Lioness website features images of the toy connected (via Bluetooth) to a phone displaying a Session Analysis – a waveform graph representing vaginal contractions – which can either be viewed in real time, or recorded and archived. Users (and their partners) can also 'watch orgasms in real time' via a moving dot on the device's app. Across the website, the device is promoted as a tool that facilitates self-improvement and communication in relationships.

In a Lioness blog post titled 'Visualise the female orgasm', the app's promoters promise an enhancement of sexual intimacy via the self-knowledge afforded by data visualisation:

> You can use live view to show a partner what's going on in the moment. It can be a very intimate and revealing experience because you and your partner are seeing yourself in a totally new way. Unlike the *Harry Meets Sally* instance of faking a real orgasm, you're showing what your real orgasm actually looks like.[9]

In her presentation at the 2020 US SexTech conference, Lioness developer Liz Klinger recounted an anecdote about how data visualisation might be used in the context of a sexual relationship. She described an occasion where a heterosexual woman was able to persuade her partner to change his sexual technique by offering him a visual proof of her increased enjoyment (implying he had not previously believed her).

These anecdotal references to the value of demonstrating sexual response via data evoke historical attempts to document the 'truth' of female bodies and orgasms, and the 'mysterious' internal nature of women's sexual pleasure (Foucault, 1990 [1978]; Williams, 1989). They also reflect the predominantly binary understandings of sex and gender that underpin most sextech marketing. Whereas the design of the Lioness website, device and app features curves and pastel colours, the Lelo Developer's Kit (a masturbation sleeve with similar data sensors) has the shiny chrome appearance of a car body, and is promoted as 'F1 high performance':

> Bridge the gap between science and nature with F1s, the next generation pleasure product for men that makes you the master of your destiny. And with a window in the casing, you can add a never-seen-before element and see your pleasure in action.[10]

Unlike the de-mystifying promise of the Lioness, the Lelo F1 masturbation sleeve is promoted to cis men as a means of engaging with data for the purpose of optimising solo sexual pleasure, enabling them to track 'progress – and prowess'.[11] Users are offered an interface that visually evokes a car dashboard – measuring thrusts, speed, pressure and temperature. While Lioness users are offered opportunities to practise altruism by donating their data, the Lelo F1 masturbation sleeve and app are sold with an accompanying software developer's kit, inviting users not simply to see and track their pleasure, but to write new code to enhance it.

The Lioness offers data visualisation and tracking as therapeutic tools and pathways to self-empowerment, and a means of contributing to a common good. The F1's design and features promise mastery, control and play with 'big boys' toys'. These marketing approaches undoubtedly build on contemporary understandings of sexual pleasure that underpin the rise of the popular feminist sex-toy market (as documented by Comella, 2017), in which masturbation is not regarded as selfish or perverse, but is de-stigmatised via its re-framing as a practice of empowerment and self-realisation. But they also resonate with historical trajectories within the fields of sex research and sexology that sought to render gender and sexuality both visible and quantifiable in normatively reductive terms.

As Annamarie Jagose (2013) observes, while the European 'marriage manuals' of the 1920s and '30s emphasised orgasm as a key aspect of heterosexual happiness, the American sexologists of the 1950s and '60s entrenched the centrality of sexual data (particularly data visualisations of physiological responses) within popular sexual imaginaries during the Global North's 'sexual revolution'. For example, William Masters and Virginia Johnson's research into human sexuality documented thousands of orgasms experienced by paid sex workers and unpaid volunteers having solo and partnered sex within laboratory settings, resulting in their theoretical model and graphic visualisation of the 'human sexual response cycle' (Masters and Johnson, 1966). Although representations vary, the human sexual response cycle (or HSRC), which maps four stages – arousal or excitement, plateau, orgasm and resolution – is often visualised as a linear graph, with the Y axis representing 'sexual excitement' – which peaks at orgasm – and the X axis representing progress over time, or duration.

This research process and its outcomes have been extensively contested by fellow sex researchers and social scientists alike. For example, Lenore Tiefer surveys a range of criticisms levelled against the HSRC – including the limitations of participant recruitment

– arguing that the diagrammatic model (and the frameworks underpinning it) is a 'self-fulfilling prophecy' based on the researchers' beliefs about normative gender and physiology (1991, p. 4). As Tiefer notes, Masters and Johnson's model is based on narrow criteria for 'successful' sexual function. The HSRC assumes penises are fully erect, vaginas self-lubricate, ejaculation is never 'rapid', and all participants in solo or partnered sexual encounters are both reliably and consistently orgasmic – indeed, Masters and Johnson excluded research subjects who did not meet these criteria.

For over two decades, the HSRC graph (or a version of it) was presented in university lectures, school sex education classrooms and mainstream sex advice columns as the norm against which healthy sexual activity could be best understood and measured. This would not be as much of an issue if the model were presented as one of *many* patterns of sexual arousal and sexual activity; the problem for Tiefer (and the HSRC's many critics) is that the researchers presented it as *the* sexual response cycle.

That the HSRC was adopted not only as the representation of a norm, but also as a normative standard of shared sexual experience, speaks to its cultural resonance. As Jagose points out, while statistical norms are often presented as objective facts, they are not always accepted as representing that which is broadly normative in a particular social or cultural context (2013, p. 50). Yet, even after the HSRC had been challenged, its graphic visualisation of orgasm remained a popular prescription/blueprint for 'successful' sexual experience well into the late twentieth century.

This brief history of sex research helps contextualise the logics (and promises) implicit in sextech marketing. Sextech's proposal that datafication is a tool for optimising pleasure, self-knowledge and intimate partner relationships is not simply an outcome of tech industry spin. It builds on professional and vernacular practices of using data visualisation and statistical analyses to make 'private' internal sensations and processes visible, generating norms and proving (or disproving) accepted 'facts' relating to sexual physiology. But while the promise of self-understanding via data cultivation builds on familiar, well-established understandings of how sex and gender can and should operate in everyday life, this does not mean that sextech cannot offer innovative or novel sources of stimulation. The Lioness website features multiple endorsements from users who frame the toy's capacity to visualise their sexual sensations not simply as a source of feedback for assessing well-being, but as a site of pleasure in and of itself. For example, a five-star review/testimonial from 'Shona B' states: 'The very first time I used the Lioness, I was blown away. Absolutely loved it! I am big on data, so I loved to see and track my orgasms. It makes me excited to have a "session." I am not just doing it to fall asleep anymore.'[12]

The mapping of physiological activities and responses (such as penile thrusts or vaginal contractions) as visual data is not simply an imposition of digital culture on an otherwise 'natural' and unmediated experience of human sexuality. On the contrary, it continues established patterns of Western discourses that seek to classify – and visualise – bodily pleasure and sexual sensation. Sextech engages with a long tradition of visualising and quantifying sex according to implicit or explicitly normative models of 'success'. At the same time, sextech devices offer new means of understanding (and taking pleasure in) sexuality as data – producing graphic representations of the intensity of physiological responses, or measuring the speed, intensity or duration of specific activities (e.g. thrusts per minute, or contractions per 'session').

Additionally, while sextech and femtech are primarily marketing terms, from a theoretical perspective they offer an opportunity to re-examine and reframe the ways that both scholarly and popular imagination has represented the interaction of bodies and technologies via concepts like cybersex and teledildonics. These terms had significant currency in early internet studies – provoking interest and debate in both the popular and tech press, as commentators speculated as to whether the technologisation of sexual encounters was a path to enhanced intimacy, a sign of emergent more-than-human cyborg sexualities or a symptom of late capitalist alienation and atomisation (Wolmark ed., 1999).

While contemporary sextech can be understood through the lens of these speculations and critiques, it also offers opportunities for collective intimacies, in which strangers share sexual experiences via collaborative data practices. In his 2019 case study of a Brazilian webcammer who configures her Lovesense Lush Bluetooth vibrator to respond to audience payments (or tips) via the Chaturbate platform, Ernesto Martins observes the ways that interactions between the sex worker/cammer and her audience are mediated by various kinds of automation and datafication. Not only does the cammer physically perform 'human sexual response' as her toy produces vibrations of specific speed and tempo in response to financial contributions, but her chatbots on the platform interface also interact with her audience via pre-programmed messages and pre-recorded sound effects. In this way, a feedback loop is created that not only links the cammer to an individual audience member, but links all of her audience members to each other as the bot allows them to observe each other's tipping practice. As Martins puts it, 'sexual activity is modulated by all participants – cammer, tipper, vibrator – and Lush plays the people who use it as much as it is played by them' (2019, p. 6). In this scenario, data flows facilitate a collective intimacy that is more analogous to the shared practices of spectatorship in a commercial sex club (or a gamer's streaming channel) than a teledildonic simulation of heterosexual intercourse.

While both sextech and femtech devices and platforms can be (and are) marketed with reference to idealised cis-heterosexual understandings of sex and gender, they offer more scope for thinking about the production, accumulation and circulation of intimate data beyond binary understandings of sex and gender. On a literal level, while user interfaces and marketing strategies are often aggressively gendered, sextech technologies themselves do not pre-determine the kinds of bodies or relationships they mediate, or the circumstances (public or private, commercial and non-commercial) in which these sexual and relational encounters take place.

Sextech and femtech also incorporate technologies that might, in other contexts, be medicalised as therapeutic devices or prostheses. These range from menstrual trackers and pelvic floor 'trainers' to assistive technologies for people recovering from childbirth, cancer and gender affirmation surgery. Sextech devices that deploy hands-free or remote interfaces can also enable people with chronic illness or disability to participate in solo or partnered sexual activity. The same marketing narratives that frame sextech in terms of self-discovery and sexual wellness can be seen as a discursive tool that facilitates the expansion of these technologies beyond the masculine sphere of porn and sex shops, and into the more mainstream (and less stigmatised) space of gift shops and celebrity influencers (Iqbal, 2021). However, these devices and platforms can be costly, both in monetary terms and in terms of the costs of exposure via data leaks.

As Flore and Pienaar observe, sextech offers developers and users alike an opportunity to build new understandings of the ways bodies, devices and data work together. As they put it, data is not simply extracted or 'mined' via sextech – rather, 'physiological data here are themselves sensory and act as lubricants of intimacy' (2020, p. 290). In her reflection on news reporting of sextech data breaches, Jenny Sundén asks 'what does it mean to have an intimate moment when connected to a device, a medium and a network that is by definition public, corporate and leaky?' (2020, p. 1). Drawing on new materialist and queer theorisations of technologies and sexuality, Sundén seeks to problematise common-sense definitions of data privacy and sensitivity. In simplified terms, Sundén suggests that the issue with sextech leaks is not so much that 'shameful' intimate data is made public, but that 'intimacy' and 'privacy' are conflated in ways that make it difficult for tech users to meaningfully consent to commercial data collection, storage and use.

Data, dating and sexual ethics

'Smart' sex toys are just one example of the ways that datafication has been framed as a vehicle to improve and enhance the legibility of intimate and sexual relationships. As we discuss below, dating app cultures offer opportunities for self-knowledge and self-expression via data (through compatibility quizzes, matching algorithms and/or the creation of a profile expressing one's tastes and preferences), with the tacit promise of intimate connection with others.

Apps shape (and sometimes limit) this capacity for connection via their interfaces and cultures of use. For example, Tinder's 'swipe logic' (David and Cambre, 2016) requires users to definitively express their interest or disinterest via a left or right swipe. However, as Bart, a twenty-one-year-old participant in Albury and McCosker's dating app research, explained, the sorting and matching logics of apps can lead to experiences of intimate *dis*connection: 'When you're talking to real people you don't normally think, oh yeah, this person is a right, this person is a left. Usually you're somewhere along the scale of between interest and disinterest' (Albury, McCosker and Evers, 2021).

As Sander De Ridder (2021) observes, apps have become naturalised in contemporary dating environments. App developers' promises to streamline connection via 'mathematisation' may lead users to expect 'predictability, controllability and convenience' in everyday sexual and romantic interactions where these elements are not, in fact, fixed or static (De Ridder, 2021, p. 5). It is unsurprising, then, that contractual consent apps – such as the Swedish app LegalFling and the Danish iConsent app – have been promoted as a one-swipe substitute for verbal negation of consent, and hence as a data-driven solution to sexual violence (Gillett et al., 2021).

Support for consent apps surfaced as part of public conversations associated with the MeToo movement, which began as a solidarity campaign for survivors of sexual violence – initiated by Black American activist Tarana Burke (North, 2019) and coordinated via the #metoo hashtag (Burgess and Baym, 2021, p. 77)

– before morphing (at the hands of white celebrities) into a more diffuse but global public conversation that saw increased attention and weight given to reports of sexual assault and harassment in the workplace, from Hollywood to Australia's Parliament House (Cave, 2021). While consent apps have been represented by developers as a means of preventing sexual assault by recording a clear record of sexual consent, critics – ranging from feminists to lawyers – have observed that not only would the apps' record be inadmissible in many legal jurisdictions, but they represent a desire to classify consent in a reductive, binary transaction (Wolf, 2018). In this context, consent is not understood as a dynamic, ongoing conversation, but as a single contractual agreement (and obligation).

For example, the LegalFling app offers sexual partners the choice of opting in or out of a range of activities, such as filming a sexual encounter or participating in BDSM.[13] It does not, however, provide an opportunity to define the details of the activities being agreed to. Within a model of consent based on 'decision tree' or 'if … then' logics, there is no opportunity to express a specific preference for some activities (e.g. taking a close-up still picture of naked buttocks) as opposed to others (taking a video of sexual activity that includes participants' faces or clearly depicts their tattoos). Similarly, clicking a button labelled with the catch-all term 'BDSM' does not allow participants to opt in or out of specific sexual practices (yes to spanking but no to bondage). While these apps may promise to mitigate against any 'ambiguity' relating to consent, they do not in fact reflect the ways that a sexual encounter (and sexual negotiation and consent) might change and evolve in real time based on subjective experiences of both sensory and affective responses.

There are, however, alternative models of consent that might better reflect the everyday ambiguities of negotiating sexual encounters within both one-off and long-term relationships. In medical research, for example, informed consent is a prerequisite for participation in research projects, including longitudinal projects that may stretch over several years (or even decades). Researchers are often challenged in these projects by the difficulty of providing clear lay explanations to potential participants, and the need to renegotiate the conditions under which data is collected and used as research practices change and evolve. In these situations, researchers may deploy purpose-built digital platforms to facilitate 'dynamic consent'.

Dynamic consent allows researchers and research participants to communicate with each other throughout a project. Researchers can update participants on their progress, and notify them of any proposed changes to research protocols. Participants are able to upload their own data directly, or renegotiate their permission for data collection and storage as their circumstances change, or in cases where they don't want to be part of new research activities or proposed third-party data-sharing agreements (Budin-Ljøsne et al., 2017). We note, too, that while the dynamic model might resolve some consent issues in research contexts, it still may not be suitable for day-to-day sexual encounters.

In settings where one partner exercises coercive control over the other, or is deliberately forcing or frightening them into sexual acquiescence, this kind of dynamic negotiation is not possible. In these cases, the scope of control can easily be expanded from forced physical activity to providing a forced indicator of 'consent' (which is not consent at all) via an app. With this in mind it is perhaps more productive not to consider whether it is possible to 'prove' consent via contractual data-collection practices, but to reflect on the ways that more pluralistic forms of sexual consent are already being negotiated, and how these negotiations might be enacted via more creative and flexible approaches to the use of data. For this reason, we turn to examples of alternative sexual cultures, where complex notions of consent are commonplace, and examine how they have adapted historically analogue practices of data collection, filtering and aggregation into digital contexts.

These models are often less dependent on 'traditional' sex/gender classifications (such as male/female or hetero/homo), and consequently tend to be more visible within alternative sexual cultures and subcultures. For example, within BDSM or kink cultures, sexual encounters between two or more partners may involve mapping a range of personal preferences for specific practices (such as rope bondage) onto preferred sexual roles (such as dominance or submission). For this reason, it is not uncommon for potential partners to share checklists that taxonomise and sort a diverse range of activities and roles according to classifiers such as levels of interest and enthusiasm.

This mode of negotiation can be conducted via the exchange of paper checklists,[14] but also can involve emailed spreadsheets, shared private links to online quiz results, or more creative data visualisations such as the Human Sex Map. The map, which invites users to place different coloured pins on sexual practices they have tried and liked, tried and didn't like, would like to try, or would prefer never to try. Here, intimate data is represented in relation to geographic terrain (i.e. the Island of Uniform Fetishes or the Land of Mundania), as opposed to more sexologically oriented checklists which are more 'diagnostic' in character.[15] Niche apps are also emerging to facilitate the kinds of specific BDSM negotiation and sexual activities outlined above. For example, the NoGrey app allows users to create and share dashboard visualisations of their sexual preferences and then overlay their dashboard with that of a potential match, creating aggregated datasets.

Digital data apps and platforms that target sexual subcultures have adopted similar strategies to allow users to filter potential matches and negotiate intimate encounters. As Ben Light (2016) – and others – have detailed, apps for 'men seeking men', such as Grindr or Squirt, not only allow users to classify themselves (and their potential matches) according to physical 'type' or 'tribe' but create dashboards that facilitate analyses of which users have demonstrated the most interest in their profile. For example, a user who has set their preference to 'Jocks' (men with an athletic build) may find that they are more likely to successfully match with self-identified 'Bears' (stocky and/or hairy men) and can adjust their search accordingly (or not) (Burgess, 2017a).

More generally, dating apps such as Tinder, Grindr, Blued and Momo offer a rich terrain for exploring the complexity of everyday data cultures. On one level, apps are supported by business models that elicit, aggregate and trade in user data (including gender, location, personal attributes, tastes, sexuality and relationship preference) (Wilken, Burgess and Albury, 2019). Like social media platforms, apps invite users to populate their profile by contributing (and frequently updating) personal data. But while users of 'real name' social platforms (such as Facebook) can negotiate workarounds by posting partial or misleading data in their profiles, this is more difficult for dating app users, given that dating and sexual connection require a degree of self-disclosure in order to ascertain mutual attraction and build trust prior to physical meetings. Many app users develop their own bespoke practices of 'two-step verification', linking other social platforms (such as Instagram or Spotify) to their app profile to share extra details of their everyday lives (and reassure potential dates that their profile is not a false identity or 'catfish') (Albury et al., 2019).

A degree of self-disclosure is required to build the trust upon which intimacy rests. But for many app users – particularly those who are not white, cis, able-bodied heterosexual men – full disclosure may put them at risk of harassment. As Bronwyn Carlson (2019) has outlined, Australian Aboriginal and Torres Strait Islander people who disclose their identities on dating apps report high levels of targeted racist abuse. Indeed, sexualised racism (which can take the form of both abuse and fetishisation) is so prevalent on dating apps that platforms including Grindr and Bumble have developed policies and campaigns to combat it (Aubrey, 2021; Mowlabocus, 2020). While platforms are yet to develop specific policies to support them, trans people are also subject to double binds where disclosing these aspects of identity makes them more vulnerable to abuse, fetishisation and exclusion, but non-disclosure leaves them open to accusations of dishonesty or deceitfulness. This is particularly apparent on platforms that offer drop-down menus that allow users diverse modes of self-identification (such as non-binary or gender-fluid), but then 'sort' all users into binary gender categories that only offer 'seeking men', 'seeking women' or 'seeking both' options (Albury et al., 2021).

Platforms that focus more on short-term hook-ups than they do on longer-term dating or 'match-making' may require users to share significant personal data, although this may vary according to the presumed gender and sexuality of app users. For example, profiles in 'men seeking men' apps may include details of HIV status, preferences regarding condom use, preferred sexual practices, fantasies and erotic roles, and the use of medications to prevent HIV transmission (Race, 2015; Albury et al., 2020). However, this level of detail is uncommon on 'women seeking women' or 'heterosexual' dating and hook-up platforms, and is certainly not an affordance of the drop-down profile menus on these apps – reflecting the stigma, risk and discomfort attached to sharing sexual health data that exists outside of gay and bisexual men's subcultures.

However, our own empirical research – as well as the work of the many scholars in the 'digital intimacies' space (Dobson et al., 2018) – has documented the multiple successful strategies deployed by people seeking sexual and romantic connections and friendships within these spaces of datification. Within our (Albury and McCosker's) study of Australian app users in dating app cultures, participants shared multiple playful examples of workarounds and 'off-label uses' (Albury and Byron, 2016; Duguay, 2020) of dating and hook-up apps that increased their sense of safety, well-being and pleasure.

For example, 'Max', a queer non-binary transmasculine person who lived in a regional city, sought to protect his personal safety by carefully limiting disclosure of his trans experience. Max described his city as a 'shallow pond', where the majority of potential matches used Tinder (as opposed to LGBTQ+ specific platforms). Consequently, he chose to game Tinder's gender-matching filters in order to be visible to LGBTQ+ people he wanted to meet, while avoiding being seen by cis heterosexuals. He described a successful strategy of logging in and out using both 'male' and 'female' profiles to filter potential matches: 'If I'm showing myself as a man I'll only look at the [queer] men, and if I'm showing myself as a woman, I'll only look at the queer women.'

Max's story is just one of many accounts of 'making do' in a limiting data environment. Other users strategically adjusted geo-locative information; adapted the pictures, texts and emoji in their profiles; or de-installed then re-installed apps to game their visibility, manipulate matching algorithms, or bypass daily limits on swipes (David and Cambre, 2016). Of course not all of these data practices are benign or playful – some are attempts to deceive or manipulate fellow users (Burgess, 2017a). We emphasise, however, that data practices relating to sexual identity and sexual desires do not need to be fully 'inclusive' or transparent in order to facilitate pleasurable and emotionally rewarding intimate connections.

The contradictions of 'careful surveillance'

So far we have considered how everyday engagements with data shape both individual and collective understandings of taste cultures, sex and gender norms, and sexual pleasures. We now turn to the equally intimate, relational space of family and everyday domestic life, in which the increased presence of data and sensors is leading to new and enhanced forms of what Ingrid Richardson, Larissa Hjorth and colleagues have dubbed 'careful surveillance' (Richardson et al., 2017) – or, alternatively, in the phrase of sociologist David Lyon (in conversation with Zygmunt Bauman), 'watching-to-care-for' (Bauman and Lyon, 2012, p. 87).

From smart door bells to apps that help us track family members' whereabouts (including those of our pets), sensor-equipped, internet-equipped and AI-enhanced smart devices and mobile apps promise increased domestic safety and enable us to watch over our loved ones at a distance, while at the same time facilitating abuse and coercion within intimate relationships. Similarly, during the COVID-19 pandemic, some states used contact-tracing apps to care for the overall well-being of their citizens – while at the same time enacting unprecedented levels of surveillance and control (Andrejevic et al., 2021).

In their book *The Smart Wife* (2020), Yolande Strengers and Jenny Kennedy situate contemporary concerns with the gendering of smart home technologies (from smart speakers like Google Home and Alexa to multi-connected surveillance and electronic control devices) against the historical backdrop of feminised domestic robots (and popular science-fiction fantasies about them). The authors highlight the tensions between the ways such technologies are in themselves feminised in their representations and functions (as domestic servants traditionally have been), and at the same time marketed primarily to men – the familiar 'boys and their toys' trope. When discourses of power and control (traditionally used in marketing technology products, mainly to men) meet discourses of care and family (traditionally used in selling products for the home, mainly to women), the lines between watchful care and coercive control can become blurred.

Tama Leaver uses the frame of 'intimate surveillance' to interrogate contemporary cultures of parenting that conflate digital surveillance and monitoring practices with care. Reflecting on emergent parenting practices – ranging from posting data and metadata associated with ultrasound images on social media (2020), to the use of digital baby-monitoring and sleep-tracking apps (2017) – Leaver observes that the monitoring and datafication of pregnancy and infancy is not only normalised, but is increasingly conflated with 'good' parenting by marketers seeking to leverage parental concerns.

In his study of marketing material associated with the Owlet baby-monitoring anklet, Leaver notes the repeated use of the term 'peace of mind', which the app's promoters directly associate with data collection and aggregation. In his reflection on how parenting influencers have amplified this association of digital tracking with 'culturally appropriate levels of care', Leaver speculates as to whether 'unplugged parenting is likely to be increasingly positioned as irresponsible and aberrant' – at least in contexts where the adoption of digital apps and devices has become widespread (Leaver, 2020).

This association of digital tracking with responsible and careful parenting is reflected in a recent study of Danish parents' attitudes towards the monitoring of their school-aged children, in which researchers found that both parents and children experienced the use of digital tracking apps as a relief from the pressure of needing to remain in constant contact as a family (Damkaer, Southerton and Albrechtslund, 2019). In these interviews, tracking was not framed as surveillance or control, but was instead understood as an expression of trust between parents and children, which alleviated the need to check up on family members' movements via frequent calls or text messages.

However, parental (and familial) monitoring is not always benevolent. An Australian study of technology-facilitated family violence found that some parents invaded their children's privacy by monitoring social media accounts or mobile devices, including installing spyware, as deliberate acts of abuse or coercive control (Dragiewicz et al., 2020). While Amazon executives have recently promoted Alexa's listening/monitoring capacity as a benign means of motivating children to do their household chores (Jargon, 2021), there is significant evidence of abusers using smart devices and other internet-connected technologies to monitor, stalk and harass their partners and ex-partners (Strengers and Kennedy, 2020; PenzeyMoog, 2021).

The line between care and control is especially blurry in marketing material that seeks to rebrand spyware as a means of ameliorating parental concerns regarding their children's digital lives. A search for 'parental monitoring apps' led us to an article titled, 'How can I track my daughter's phone without her knowing?' The article explained how to use common apps such as Google Maps and Find My Friends as spyware, and recommended using the Secureteen service to covertly monitor 'chats, call logs, filtering apps and websites' in the name of child protection. Once parents have subscribed to the service and accessed their child's device, they are given a dashboard that allows them to access the child's apps and monitor their keystrokes.[16] The implicit premise of these apps – that data surveillance equals protection – points to the paradoxical position of young people within data cultures, and the risk, shame and anxiety attached to them.

Cyber-safety messaging encourages children and young people to be vigilant against 'strangers' online, guarding personal information such as their surname, school, age and location. Young people are not only warned against sharing intimate data with strangers – their friends and schoolmates are also risks to data privacy. As feminist researchers have noted, 'sext education' focused on preventing image-based abuse cautions young women to mistrust both their school friends and crushes/lovers, with warnings that unwanted data-sharing is inevitable (Dobson and Ringrose, 2016). Albury and Crawford (2012) observe that this style of data-safety advice evokes traditional forms of sexual violence prevention, absolving perpetrators and framing young women as wholly responsible for preventing abuse via the vigilant management of their personal data.

Prior to the era of internet-enabled phones, parents were encouraged to monitor their children's online activities by insisting that personal computers were only used in shared spaces (such as living rooms and kitchens) as opposed to bedrooms. With the advent of social media platforms and mobile devices, parents were encouraged to insist that children shared passwords with them, and to limit opportunities to browse, share pictures or chat 'inappropriately'. These forms of monitoring were framed as expressions of loving care, but as danah boyd (2012) observed in a blog post, the framing sent mixed messages to both parents and young people alike. On one level, young people are told that a desire for digital privacy is equivalent to 'keeping secrets' from their parents, and that parental surveillance is an expression of love and trust; on another level, they are told that their close friends or romantic partners can *never* be trusted with personal data, no matter how much they love or care about each other.

This tension between disclosure and privacy in data imaginaries does not just impact on young people. In a project developing a hands-on model for improving digital literacy, a group of older participants (average age seventy-three) explained the core elements of an underlying data anxiety (McCosker et al., 2018; Bossio and McCosker, 2021). The most common anxiety or difficulty with social media use involved *personal data control*. Whereas young people's data anxieties focus on location and whereabouts, the seniors' strongest concerns were about financial information and security (Pangrazio and Selwyn, 2018). Personal information and data were understood by the group mainly as a means of controlling access to personal assets. One account captured the tension behind their reluctance to engage actively online: 'It's just the fact that on the one hand I believe in total openness, and on the other hand I believe in being a protector of data. There's not really a logic to it. It's an instinctive thing. I just say, why do I need to reveal that information about myself?' (Henry). These data anxieties are then extended to anxieties around control over interactions:

> [It's] not the fear of social media. It's the fear of the implications of the handling of social media. That I might say something that I wish to be said in private, but other people repeat it. I know you can go private and things like that. But you have no control over what other people may want to repeat. And this is a fear of mine. (Rumbles)

Many of the participants suggested they had undertaken careful boundary work around the exposure of personal or intimate details through profiles and posts, and remained sceptical of those who they felt gave too much information about their lives. They felt that Facebook was too public, too open, despite the offer of privacy controls, revealing a generational disparity in expectations about exactly how we might be visible and what pieces of personal data are available to others.

Conclusion

In this chapter we have explored the ways apps and platforms come to know us as we generate, curate and share intimate data. We have looked at sextech start-ups and their desire to promote various forms of data-driven intimacy through the creation of devices and data dashboards. We have investigated dating app cultures, where the extraction and manipulation of data is both 'baked into' apps (as a requirement of their use) and actively performed by app users themselves as they seek to connect with potential partners. And we have looked at 'intimate' or 'careful' surveillance that promises increased domestic security and familial intimacy, while at the same time facilitating abuse and coercion within some intimate relationships.

The idea of everyday data intimacies, then, is about how data has become bound up with the ways we know ourselves, and how it has come to mediate the process by which we come to know and trust each other (especially but not only in close, dyadic relationships). Data, too, provides the means through which an intimate knowledge of people in our lives can be converted into control or abuse of these relationships.

All these examples embody the tension in intimacy identified at the outset of this chapter, where the inwardness of intimacy pushes up against a corresponding publicness. We see this, for instance, in the platform interplay between knowing ourselves through data and the algorithmic encouragement to circulate and share data and participate in collective intimacies. Apps, platforms and start-ups might position themselves as proffering 'solutions' to the complications of everyday intimacies. Yet what this chapter also reveals is that the nuances and complexities of everyday intimacies only serve to complicate these efforts. Everyday intimacy remains a site of complex negotiation and contestation, and the push for 'networked' or 'data-driven' intimacy raises some difficult ethical challenges.

In the following chapter, we explore data literacy as another dimension of how people interact with everyday data (whether intimate or otherwise), and as an important site of data's everyday politics. Grounding the discussion in the cultural studies tradition of scholarship around literacy's historical, cultural and political dimensions, we identify and examine the tactics, strategies and capacities through which we make over and make do in our personal encounters with data.

4

Everyday Data Literacies

Data has traditionally been the domain of the mathematicians, computer scientists and engineers – for the rest of us, perhaps, it has carried an air of elusiveness as well as a sense of gendered (masculine) prestige and authority. The numbers and complexity of big data can seem impenetrable. But, arguably, we all now have a stake in understanding and making use of the data that surrounds us.

This chapter explores the ways that, despite its complexities, people come to know about and manage to do things with the data that runs through their everyday lives. Given that the everyday data cultures framework trains our attention to how people live with, through and as subjects of data, this chapter considers the kinds of skills, capabilities and new forms of data expertise with which they do so. Adding another layer to our framework, we examine some exemplary forms of vernacular data expertise that emerge in response to daily interactions with technology. We draw on and expand the concept of data literacies to take a closer look at the work people do with data and how they respond to the data-driven feedback loops that shape digital systems.

First, what is everyday data literacy, and why is it important now? There's a broader context. Literacy and education have for some time been considered key sites of social and cultural struggle (Freire, 1968; Hoggart, 1957). Literacy is understood as both formal (school) and informal (everyday), but is also historically constructed, involving knowledge of systems of reading and writing, as well as the skills needed to work within those systems. Considering encounters with data or data-driven systems – in the form of awareness, understanding and use – through the lens of literacy helps to uncover the way it is shaped and contested from the ground up.

We're often told now that getting ahead in the data economy, or even just getting by, increasingly means building personal data capability – knowledge, expertise and skill. Microsoft boasts that one in five adults worldwide now uses Excel (Hubel, 2017). Corporate data literacy courseware and professional development programmes are on the rise. While a recent UK report shows signs of a drop in computer science participation at high school and university, the proliferation of DIY online learning and other coding and data science toolkits is clear to see (Learning and Work Institute, 2021). Data literacy, the story goes, is for everyone. But we can also ask: 'data literacies to what ends, and in whose interests?'

While data literacy can certainly conjure limited or complicit ideas of corporate power, there is plenty of room for a better understanding of the way people learn about and respond to the logics, infrastructures and flows of data under widening conditions of datafication.

D'Ignazio and Bhargava (2015) define data literacy as 'the ability to read, work with, analyse and argue with data', emphasising both the technical and communicative competencies. This has been extended to include critical awareness and navigation of the role data and data-driven systems and infrastructures play in our personal and social lives (McCosker, 2017a; Gray, Gerlitz and Bounegru, 2018; Pangrazio and Selwyn, 2019). Data literacies are plural (literacies) even if we often refer to 'literacy' in the abstract singular. By targeting *everyday* data literacies, we highlight the ways that, in the course of daily life, vernacular knowledges and skills intersect and conflict with institutional imperatives.

In what follows, we observe everyday data literacies in the use of technology and systems for work, personal health and well-being, as well as social interaction and entertainment. Data literacy is a concern for gig-economy workers trying to make ends meet, and for warehouse and delivery workers engaging with monitoring systems to meet hourly quotas. Data outputs guide the way we use devices and apps to support decisions about our personal productivity, health and well-being. And some skill with data inputs and algorithmic calculations is useful for anyone attempting to navigate and find some sort of visibility and success in social media or service platforms from eBay to YouTube or TikTok.

These situations require vernacular skills, knowledge and tactics that operate within, alongside or against formal structures. Scholars investigating how YouTubers and TikTokers navigate the recommendation algorithm refer to this as 'gaming the system' (Petre et al., 2019) or 'playing the visibility game' (Cotter, 2019), and describe the 'folk theories' that both react to and shape data-driven systems (Ytre-Arne and Moe, 2021).

More broadly, we're talking about the kinds of tactics, 'tricks and competitions' that de Certeau described in relation to people getting by and finding benefit in work environments that also impose relationships of subordination (1984, p. 29).

Examining the way people use everyday data literacies to make do with data, and to make something of the systems and platforms built on that data, offers a lens through which to understand platform and technology use as sites of struggle. In this chapter we tell some of these stories of everyday data literacy in practice.

The uses of data literacy

Data literacy may appear to be primarily about numeracy, relating to formal data science or the mastery of spreadsheets, but it is much more than that, given the role played by data in all our interactions with automated and algorithmic systems. It has become an increasingly important set of capabilities for negotiating modern life, along with digital, computational, information and media literacy, for example (Koltay, 2011). It can be defined narrowly as the ability to read, analyse and work with data (D'Ignazio and Bhargava, 2015). But it has been attributed to a wider range of critical data practices, including the ability to identify and navigate personal data collections and settings (Pangrazio and Selwyn, 2019), platform metrics and analytics (McCosker, 2017a), data infrastructures (Gray et al., 2018) and algorithmic systems (Gran, Booth and Bucher, 2021). While some have drawn distinctions between data literacy and the notion of expertise (Bassett, Fotopoulou and Howland, 2015), we see these as related and similarly contested aspects of the challenges facing diverse people in an age of datafication and automation. At its base, the idea of data literacy concerns the ability to identify, read, critically analyse and act in relation to data in our personal environments, while the idea of data expertise emphasises certain claims to 'mastery' and authority associated with these capabilities.

These perspectives follow a long history of scholarship illustrating and advocating the role of often informal and culturally situated modes of learning as a way of addressing social problems. In Hoggart's classic early cultural studies work (1957), as in Street's (1984) 'new literacy' and Freire's critical pedagogy (1968), literacies are at the front line of individual empowerment and collective political struggle. One of Hoggart's lessons in his *The Uses of Literacy* was that, despite the assumed effect of the then-new mass media, working-class life carried on in its rich traditions of language and speech. Lawrence Grossberg argues that Hoggart's principal contribution here was 'to give expression to a deeply humane and humanistic sense of moral, intellectual and political possibility grounded in the lived realities of ordinary people' – not in the abstract, but in 'the concrete processes of cultural and social transformation'. For people to be 'agents rather than objects of historical change', though, 'ideas, knowledge and critical understanding were necessary' (Grossberg, 2015, p. 106).

During the 1950s and '60s, the new mass media ecosystem of publicists, pulp novels, magazines, and popular music and film transformed what it meant to be literate; but likewise, 'reading' popular culture was not just a matter of taste and education. New modes of reading and communication emerged. Fiske and Hartley (1978) demonstrated the historical and technological contingency of literacies by detailing television's temporal and 'bardic' logics – that is, the continuous flow that makes television more like speech than like writing or literature. In this sense literacies are resources for public connection and belonging. They are not only individual skills for succeeding at work or school, but commonly practised ways of engaging with media and information. Media literacy emerged out of the subsequent push in education to equip students to respond critically and creatively to the mass broadcast media environment (Buckingham, 2003; Sefton-Green, 1998), as did information literacy in the context of increasingly complex library, information and internet search systems (Koltay, 2011). A proliferation of 'adjectival literacies' (Watkins, 2015) has followed with each new wave of media change.

Rather than static competencies that individuals develop or possess, then, literacies are dynamic, situated historically, politically and socially, and are fundamental to driving change in new cultural and technological milieus (Ong, 2013; Barton, Hamilton and Ivanič, 2000). Importantly, the 'new literacy' movement of the 1980s and 1990s shifted the focus from individual cognitive ability to collective competencies – the dynamic, situated, relational and cultural ways of using language, making meaning, and acting in the world. As an umbrella concept, digital literacies have been commonly understood in terms of education and empowerment, a response to ever changing digital environments (Gilster, 1997; Knobel and Lankshear, 2008; Pangrazio, 2016). Jones and Hafner (2012) see digital literacies as an adaptive set of abilities, skills and knowledges about the operation, use and cultures of digital media and networked systems. Accordingly, mastery of digital platforms like YouTube, for instance, means 'not only being able to create and consume video content, but also being able to comprehend the way YouTube works as a set of technologies and as a social network' (Burgess and Green, 2009, p. 72).

The idea of critical data literacy is widely understood as a response to the challenges that datafication poses to social life and individual agency (McCosker, 2017a; Pangrazio and Sefton-Green, 2020). The 'opacity' of data collection and flows, and the operation of computational and machine learning algorithms (Burrell, 2016), have attracted substantial critique. But they have also encouraged new strategies for lay learning, disrupting the divisions that cultivate 'digital expertise for the purposes of industry, whilst refusing to make it a demand for everyday life' (Bassett, Fotopoulou and Howland, 2015, p. 337). Alongside the hype and enthusiasm for data science as a way of generating corporate advantage, data literacy carries associations with self-help and professional development. This has led to a proliferation of commercial courses and learning resources. At the same time, and perhaps in response, critical approaches have been devised with advocacy and activist goals. These approaches look to ways people can actively work with and respond to the data produced through interactions with platforms and devices. Often the goal is understanding and managing personal data – the data that identifies and comes to define us. A personal data literacy framework begins with an awareness and understanding of the processes of data capture, developing reflexivity and an ability to respond to the technologies generating personal data, and a higher-level ability for tactical response (Pangrazio and Selwyn, 2019).

Everyday data literacy can be understood as a new battleground for managing and mastering platforms, apps, tools and systems that encompass our health and well-being, working lives, and cultural and social activities. Data literacies are often at play in workers' interactions with workplace productivity metrics, as well as in the skilful self-accounting of personal health and wellness tracking. They also emerge through the process of learning about, negotiating and attempting to master the algorithmic systems in use across social media platforms. We explore each of these sites of data struggle and vernacular expertise in turn.

Working with data

The platforms created to make white-collar work more efficient have already laid the foundation for new forms of workplace measurement – think email, Excel and the rest of the Office suite, as well as connective technologies like video conferencing, or integrated computing and cloud services. Microsoft's 'smart assistant' Cortana ingests data across Office apps – 'reading' our email, noting our temporal patterns of app use and suggesting when we should be blocking out some 'focused time', nudging us when we're late for a meeting or have forgotten to send a file, and serving up a whole range of productivity stats judging the week's activity and outputs.

Similarly, software systems for factories and manufacturing operations have established sophisticated forms of control through counting and measuring activity, stock and inventory, process flow and maintenance needs. Corporate software and monitoring systems are rendering the 'digital exhaust' emitted by workplace activities as data and analytic insights. Productivity platforms and apps also operate as a space for cultivating self-care and healthiness. Microsoft Viva Insights, for example, seeks to apply AI to work activity data through word processing, spreadsheets and email to 'help people nurture wellbeing' and be their 'best self' (Patton, 2021).

These and other forms of workplace datafication raise questions about the impact on workers and the level of everyday surveillance we're willing to tolerate as a society. But the questions that interest us here are about the conditions and competencies of everyday data use on behalf of workers, and the increasing need for data literacy as a basic function of working life in the face of these intensifying organisational dynamics of datafication. What are the new forms of knowledge and data skill that are needed to cope with, succeed in the face of, or push back against, these productivity measures, metrics and algorithmic systems? These are not just questions for white-collar workers. They are also highly relevant to those navigating the 'gig economy' – the kinds of work mediated by digital platforms and data infrastructures forged by Uber, TaskRabbit, DoorDash or Airtasker. And they are at the heart of the concerns of workers seeking decent working conditions within automated systems under the punishing hourly and daily targets imposed by Amazon's warehouse and delivery work.

The century-old science of productivity and optimisation feeds off the data we generate as part of everyday work practices. The platforms that support that work are designed with measurement and accountability in mind. Scholarship exploring the datafication of work, or 'efficiency engineering' as Melissa Gregg puts it, describes productivity as the staple measure used to calculate the outputs of daily work – the data and metrics that weigh levels of efficiency and the outcomes of labour (both for the organisation and for the individual). While platforms and systems are clearly imposed on us as a condition of employment in different industries, the productivity they seek to shape is also 'experienced as an archly personal, everyday concern' (Gregg, 2018, p. 3). Competition, precarity and changing work environments require us all to constantly capture, measure and prove our worth.

The imperative to 'do more with less' and 'work smarter, not harder', Gregg reminds us, has become the internalised call of productivity measurement, and underpins an entire industry of 'productivity support' devices, platforms and apps. Productivity tools aim to turn data generated as a by-product of everyday work practices into insights and efficiency. This level of skilful self- and work-optimisation might be celebrated as the pinnacle of data-driven office work, but these competencies introduce new workplace divides. Van Dijck, Poell and de Waal (2018) have detailed how digital and data-driven platforms gradually infiltrate and converge with the institutions and markets around which contemporary societies are organised. Urban transport, health and fitness, education and news each attract large-scale tech industry investment in new platform infrastructures that both produce and depend on data flows. These developments are accompanied by new platformed and data-driven forms of gig-based work, characterised by platform-mediated transactions between service providers and customers.

Just as gig-economy companies like Uber have altered employment structures to see workers as at best independent contractors, those workers are also left to 'work the system' on their own terms (Woodcock and Graham, 2019). They have to become adept at managing digital and data systems in search of a fee for their service. But despite the isolating tendencies of platform labour, gig workers find each other, and they talk: animated backchannel discussions circulate among workers hoping to find better ways of scraping a living from the micro-payments of the digital economy, and sharing hard-won tactical knowledge. A driver seeking jobs might be juggling multiple apps, checking routes for congestion on Google Maps, looking for events that might generate demand, or times of the day when people are most likely looking for rides. The work requires heightened skills and tactical insight into how to hack the quotas, charges and frequency of 'ride pings', literally navigating the edges of the technology-imposed rules of taxi work. Discussion on forums like UberPeople.net weigh the benefits of running multiple ride-sharing apps simultaneously – Uber, Ola, DiDi and others. One contributor to the forum calculates the benefit as follows:

> But if there are 10 pax [people] looking for a ride, and 6 of them are using Uber, 2 DiDi and 2 OLA, then you are only able to get 60% of the ride market if just using Uber. When pings are hard to come by, one needs 100% of the market. (D-River, 2019)

The gig workers who participate in these discussions game the system, and share their knowledge, in an effort to do more with less in platformed and datafied work environments where the parameters are constantly changing, and where they are positioned in competition with each other. Their experiences echo a set of workplace practices that Michel de Certeau (1984) discussed in the context of the cubicles of the 1980s corporate firm. He used the idiomatic French term *la perruque*, a socio-cultural trope, to describe situations where those who work for others in menial jobs, for instance in low-paid roles in large companies, make use of their situation to carve out some personal benefits. This might include doing some personal photocopying, planning a holiday, paying bills, or, more recently, browsing and scrolling through social media. These disruptions to the routines and counter-routines represent the kinds of 'tactics' that challenge the all-encompassing impositions of workplace task managers.

To some extent, the datafication of workplace trans-actions and services flips the salary-time efficiency model to capture, quantify and value only the point of transaction, and cut loose the waiting or downtime. Uber's tagline of 'be your own boss' carries substantial implications for the work contract (Woodcock, 2020). Years of court battles and regulatory rulings have hinged on drivers' categorisation as contractors or employees. The status has implications in the United States, for example, for federal and state tax, healthcare, anti-discrimination, worker's compensation and unemployment insurance obligations (Wood, 2020). The 'promise' of gig-economy work is that workers are free to determine their own 'jobs' as they arise in a system built on transactional data. The result is that non-productive activity no longer registers as part of the work-day contract.

Navigating the data arrangements established through a gig-work platform becomes a point of focus in the attempt to balance benefit and labour. DoorDash deliv-erers have to calculate and weigh up delivery distance, fuel consumption, direction and traffic when deciding whether to take a job. There are reports of 'dashers' working together 'to turn down the lowest-paying deliv-eries so the automated system for matching jobs with drivers will respond by raising pay rates' (Ford, 2021).

These technical structures do not necessarily entail only one-way exploitation. Research exploring the experiences of workers in Africa show the complex negotiations and trade-offs associated with the agency and restrictions that gig-work platforms introduce. Anwar and Graham's study of workers using the Upwork digital work platform in South Africa, Kenya, Nigeria, Ghana and Uganda draws out some of these negotiations. They show that remote workers in these countries manage constraints on Upwork through 'diverse everyday resilience, reworking and resistance practices' (2020, p. 1269). The situation is problematised in cultural industries, with workers navigating platform metrics to achieve visibility and get by or get ahead, further disrupting the boundaries between everyday life and labour (Duffy, 2018), as we show later in this chapter.

Datafication can invite new forms of data literacy, but it can also diminish others. In the folklore of London cabs, 'The Knowledge' was a taxi driver's internalised knowledge of London streets and routes. An effective black-cab driver had to study the roads and know every potential destination and the optimal route to get there in order to pass the gruelling licence test. Often, years were spent taking notes and testing routes at different times of the day. 'A Knowledge student can commit to memory not just the streets but the streetscape – the curve of the road, the pharmacy on the corner, the mice nibbling on cheese in the architrave' (Rosen, 2014). With GPS, of course, and the real-time traffic-flow data processing available through Google Maps or Waze, that knowledge set has been challenged. Now, calculations of the number of turns, potential traffic jams, direction and speed are far more likely to be delegated to an app.

Within this broad scenario of the datafied and platformed society, cultural *and* data competencies play an equally important role in how work is being reconfigured. In the D-River example above, there is some sense of agency, or perhaps a vernacular form of applied statistics in the calculations of the ride market, as drivers juggle and work the systems of multiple apps. GridWise is an app built to help drivers navigate exactly this kind of scenario (Weed, 2019). It draws on additional datasets to generate a real-time lay of the land to help them calculate the best routes, the most profitable rides to take on, the best pick-up locations and times of day and so on. Of course, like other apps of its kind, GridWise has the advantage of employing teams of data scientists to compile and process relevant data from across digital sources. It's a reminder that the data competencies needed to navigate the digital economy are not evenly distributed; and as the displacement of The Knowledge shows, hard-won vernacular literacies developed through everyday experience remain in tension with the constantly shifting parameters and constraints of institutional systems.

Self-tracking

The data logic of self-optimisation goes well beyond labour productivity in the workplace. Personal health tracking has become a major device- and platform-supported industry. It feeds into and encourages certain kinds of cultures of data collection, sharing and use for personal gain. A growing body of research has sought to understand how self-tracking and health data visualisation has created new kinds of awareness and control over personal health and well-being (Neff and

Nafus, 2016; Swan, 2012; Lupton, 2016; Lupton and Smith, 2018; Jethani, 2021). However, we'll contrast this with a grittier picture of the struggles over personal health and activity data in the ruthless tracking of work patterns and 'targets' in the high-tech, high-stakes warehouses of Amazon's product distribution empire. At the – so far – extreme edge of bio-monitoring and activity management through biosensor data, Amazon has proposed algorithmic warehouse arrangements where workers are shifted from one activity to another on the basis of musculoskeletal distribution tracking data, as a form of personal quantified self-care. This and more consumer-led forms of health data tracking paint a picture of data literacy as an uneven and unsettled site of struggle.

The tools for self-tracking proliferated from the early 2010s with developments in sensors combined with wireless data processing. Data tracking devices like Fitbit and Garmin watches and bands made their earliest inroads into everyday life through health and wellness tracking. The so-called 'Quantified Self' movement emerged around that time, but, as many have noted, this is only a more recent manifestation of a long history of self-tracking and personal data analysis (e.g. Humphreys, 2018). Neff and Nafus detail the contested senses of the idea of the quantified self. While the idea came to refer to a more or less cohesive community of enthusiasts, there are many contexts in which 'people wrestle with what data means', especially what it means or can do for their health and well-being (Neff and Nafus, 2016, p. 34). We referred to this in Chapter 2 as a search for granular certainty through compositions of the 'statistical body'.

Self-optimisation and health self-tracking have contributed substantially to revaluating everyday activities, routines and events in the form of data. In an early analysis of wearable self-tracking devices and the networked sensors that support them, Swan observed the 'new data literacy behaviours such as correlation assessment, anomaly detection, and high-frequency data processing' that emerge in response to self-tracking, 'as humans adapt to the different kinds of data flows' these devices and applications and interfaces offer (2012, p. 217). But the 'data accounting' associated with these technologies may be more mundane than this (Humphreys, 2018). A 2013 study in the United States found that 60 per cent of adults track a health indicator like weight, diet, exercise routine or a health symptom (Fox and Duggan, 2013). The authors also note that many track 'in their head' or on paper as well as through health apps and devices.

The early enthusiasm for Fitbits and other tracking bands may have waned with the mass take-up of smart watches, but contemporary ideas about self-care and well-being are as infused with data as ever – and, more generally, we are increasingly asked to participate in data-producing digital health practices as a part of datafied healthcare services. But the ongoing datafication of health monitoring meets with just as much resistance as enthusiasm. Some researchers locate this resistance in people's awareness of the gendered demands placed on users by self-tracking technologies, which police women's bodies in particular (Esmonde, 2019). Others attribute ex-users' abandonment of self-tracking devices to data uselessness and measurement inaccuracy (Attig and Franke, 2020); it also takes not only skill but time and effort to get beyond the generic data dashboards that come with Strava, Fitbit and similar health and fitness trackers.

For some people, though, the availability of health data tracking tools gives rise to experimentation with the 'explanatory power' of health and well-being data. For example, the sub-Reddit r/DataIsBeautiful offers many examples of self-tracking, bio-hacking and personal optimisation in the form of health data visualisations. As Robards, Lyall and Moran show, personal data on r/DataIsBeautiful becomes the acceptable medium through which the primarily younger men of the forum examine and publicly share their intimate side with others. With their self-oriented, confessional edge, many of the data visualisations collected in this popular space capture and re-present versions of an everyday self, from, for instance, 'dating ("My 500 days on OkCupid") to health data ("My health decline and eventual brain surgery") through to incredibly detailed lifelogging ("Every Single Hour of My 2017 Recorded")' (Robards, Lyall and Moran, 2020, p. 2). And these activities exemplify the diffusion of health data expertise.

At a broader population level, personal health and well-being data tracking has transformed everyday participation in healthcare processes, and contributed to shifting forms of knowledge and expertise. Ruckenstein and Schüll's (2017) research shows that personal health data is increasingly embedded in everyday meaning-making practices, and is no longer the objective immutable register of 'the' health status of populations, people, bodies and behaviours. Nor is it solely the domain of medical and health specialists. At least in theory, digitally included people have more access to health information and their own health data than ever before. Fiore-Gartland and Neff (2015) explain the different values and interpretations of health data among different groups such as technology designers, medical practitioners, patients and self-trackers. They show how in some circumstances personal health data presents a kind of 'self-evidence', or actionable information. Through the availability of personal health data, new forms of expertise and health literacy are forming, shifting the relations of care and professional practice in this domain.

But the everyday data practice of self-tracking we have described above is only one model of health data tracking, and one associated with relative privilege. An entirely different story emerges around the tracking of health data in the workplace.

In his final letter to shareholders as CEO of Amazon before stepping down to become Chairman of the Board, Jeff Bezos expressed his disappointment with ongoing patterns of poor worker well-being and high frequencies of workplace injury. As one of the largest private employers in the United States, the stakes are high for the everyday welfare of Amazon's warehouse workers and delivery drivers. As noted in the Introduction, Amazon has some of the most high-tech, robot and data-driven warehouse systems in the world. However, a spotlight has been shone on the damage done to its immense workforce by the company's gruelling productivity targets. Accounts have circulated of delivery drivers peeing in bottles to save the fifteen minutes off-route a bathroom break would add (Vincent, 2021c). But Bezos's statement about his ambitions to address workers' conditions with an algorithmic fix has raised an interesting conundrum concerning the place of data in designing workplace well-being. Bezos pronounced that:

> We're developing new automated staffing schedules that use sophisticated algorithms to rotate employees among jobs that use different muscle-tendon groups to decrease

repetitive motion and help protect employees from MSD (musculoskeletal disease) risks. (Bezos, quoted in Ongweso, 2021)

The solution to improving workers' well-being, it seems, is to increase capture of biodata to better automate activity distribution about the warehouse 'based on which isolated muscle-tendon groups they're repetitively grinding' (Ongweso, 2021). To backtrack a few steps, Amazon workers are well aware of the punishing effects of 'the rate' on their bodies. They know the counts and marks that need to be hit to avoid warnings or losing their job.

> Dixon's scan rate – more than 300 items an hour, thousands of individual products a day – was being tracked constantly, the data flowing to managers in real time, then crunched by a proprietary software system called ADAPT. She knew, like the thousands of other workers there, that if she didn't hit her target speed, she would be written up, and if she didn't improve, she eventually would be fired. (Ongweso, 2021)

Similarly, Evans recounts the experience of a disabled veteran who worked at the Troutdale, Oregon, warehouse but fell afoul of ADAPT:

> The expectations were precise. He had to pick 385 small items or 350 medium items each hour. One week, he was hitting 98.45 percent of his expected rate, but that wasn't good enough. That 1.55 percent speed shortfall earned him his final written warning – the last one before termination. (Evans, 2019)

Reports of serious injury logs at different warehouses build aggregates of individual data stories like these.

The Troutdale warehouse had rates of serious injury as high as twenty-six per 100 workers in a year, and in many warehouses the figure falls between five and fifteen in 100 workers (Evans, 2019). These and similar data have been used (so far unsuccessfully) to push for the unionisation of Amazon warehouse and delivery workers. There is a sense that the extensive systems of worker surveillance and labour datafication can have their uses beyond identifying the human breaking points and how they can be avoided by the company. They also introduce new ways to identify precise forms of often hidden workplace exploitation and injustice.

The larger point of contention is about how, in all of this sophisticated data analysis, worker autonomy might be improved rather than diminished, and how data can be used to truly improve workplace well-being rather than function as a proxy for low productivity or factory-machine breakdown, as in Bezos's attempts to shuffle workers about on the basis of muscle-tendon-use data. The difference here is about whether an employee at a data-intensive workplace is positioned to work *for* the automated machinery, contributing to its capability for productivity, or whether the machines are designed to help and augment workers' activities, improving their health rather than diminishing it. Either way, with more personal well-being and health-status data available than ever, the need for improved data competencies is becoming vital.

In many critiques, self-trackers are depicted as parts of the capitalist machine, with workers 'enacting cultural values of entrepreneurial, autonomous behaviour, responsibly managing and optimising their lives as cogs in a neoliberal wheel' (Schüll, 2021, p. 460). The equation of self-tracking and dataveillance here rings true in regard to the valid criticism of

Amazon's hyper-surveillant workplaces. Self-tracking, though, has shown that in certain circumstances, with a little knowledge and ability, ordinary people can own, control, make their own sense of and benefit from their personal data. Reflecting on the rise and dissipation of self-tracking as a movement in recent times, Schüll argues that 'the power of self data lay in the relational grammar that emerged across their data points' in the storytelling and activation of data as a mechanism for change (2021, p. 463). But we don't necessarily have the mechanisms and tools, or the shared language – the literacies and expertise – to manage these connections. The following section explores the ways people are developing such a shared language of the operations of data and algorithms across social media platforms.

Algorithm literacies

One of the more notable developments in the rise of TikTok as a social platform is the collective effort to co-learn and affect its notorious recommendation algorithm. On TikTok, visibility and attention are the main business. Driving it is a complex and ostensibly hidden system of automated decision-making fed by search, interaction and engagement activity. A growing collection of empirical research is showing the way ordinary users across various platforms understand and negotiate this data production, and how they seek to 'get on top' of social media algorithms (Bucher, 2018) as forms of collective literacy and peer pedagogy. As a final set of examples of everyday data cultures involving increasingly active and sophisticated data literacies, we turn here to the algorithms of popular social media platforms.

Everyday data literacy practices can be found in the way social media users drive and respond to 'like' counts, interact with content, devise a profile or manage search strategies. These acts are generative of data and contribute to the shaping of algorithms. A raft of research has begun to explore these characteristics of algorithm literacy. There is growing evidence showing how 'successful' use of social media platforms involves skilfulness and expertise in 'playing the visibility game' (Cotter, 2019; Bucher, 2018). For Cotter, the 'pursuit of influence resembles a game constructed around "rules" encoded in algorithms', setting up an inter-dependence between users, algorithms and platform designers (2019, p. 895). This pursuit involves the development of platform data literacies, or an ability to master the visible metrics and analytics available through platforms, apps and devices, developing tactics for reading and negotiating them to gain platform advantage (McCosker, 2017a).

If algorithmic recommender and ranking systems are driven by data inputs from their users, to what extent are those users able to individually or collectively shape, manipulate or game those systems? Because the workings of the algorithms and the data inputs remain either hidden or complex, some researchers have sought to understand their operation through users' own folk theories and collective action (Siles et al., 2020; Ytre-Arne and Moe, 2021). Folk expertise relies on knowledge about the key signals that affect a platform's visibility or profiling metrics, or the weighting applied to kinds of content and rates of engagement (McCosker, 2017a). For Bishop, signif-icant technical expertise is traded among beauty vloggers on YouTube: 'their strategic management of algorithmic visibility makes them an illuminating source of algorithmic knowledge' (2019, p. 2589). Algorithmic gossip is the term she uses to explain the social learning involved in developing the expertise required for success on YouTube. These are the 'communally and socially informed theories and strategies pertaining to recommender algorithms, shared and implemented to engender financial consistency and visibility on algorithmically structured social media platforms' (2019, p. 2602).

Social media platforms offer generative case studies for observing and understanding data practices as forms of algorithmic literacy. Referring to the concept of A/B testing in software systems (real-time experiments testing two product design options on two groups of users), van Dijck, Poell and de Waal remind us that 'any platform is a recalibration laboratory where new features are constantly tested on users' (2018, p. 11; citing Benbunan-Fich, 2017). But with the curation and recommender algorithms driving contemporary social media platforms, to what extent does it actually 'take two to tango', as Nick Clegg (2021) puts it?

In a provocative article posted to *Medium*, Clegg – Facebook's Vice-President of Global Affairs, former deputy prime minister of the United Kingdom and leader of the Liberal Democrats – weighed in on the issue of how much control and agency people have in their engagement with Facebook and its personalisation and ranking algorithm. He set out to counter what he sees as the growing 'doomsday' critiques of Facebook's algorithmic manipulations. Ultimately, he bluntly states, 'content ranking is a dynamic partnership between people and algorithms. On Facebook, it takes two to tango.' On Facebook's behalf he concedes that social media companies 'need to be frank about how the relationship between you and their major algorithms

really works. And they need to give you more control.'
His solution to this problem, however, is both telling
and problematic. It misunderstands some key aspects
of what data and algorithm literacies are. Not least,
Clegg's account assumes literacy is solely an attribute
of individuals, overlooking the situated, contested,
collective, interpersonal and relational literacies we
have discussed in this chapter.

Under the header, 'How to train your algorithm',
Clegg provides some strategies and tips on how people's
choices and the ranking algorithm work together in the
newsfeed: 'Thousands of signals are assessed [for each
post], like who posted it, when, whether it's a photo,
video or link, how popular it is on the platform, or
the type of device you are using.' These are the core
action and interaction data points that Facebook is
most interested in – both for commercial reasons and
to predict what users want to see or see more of. He
goes on to explain that the algorithm uses those signals
to predict how likely a post is to be meaningful to you.
Actually, it's predicting whether you will 'like' the post,
'or find that viewing it was worth your time'. But there
are other layers of filtering going on, and new layers
(and live experiments on users) being introduced all
the time.

Facebook has previously adjusted the signals to
promote content from family, friends and groups, and
to demote pages and content from companies and news
organisations (it now makes these 'pay to play'). And
Clegg notes that the platform is considering how to
rank some important categories like news, politics and
health differently. How is an ordinary user expected
to 'tango' with this complex and constantly changing
algorithmic set-up? According to Clegg, it's by feeding
the system *additional* data and information through

a range of platform tools and toggles. In a strikingly similar vein, Instagram boss Adam Mosseri posted a video explaining Instagram's sorting algorithms, and repeating Clegg's rationale: 'It's hard for people to trust what they don't understand, which is why we wanted to shed more light on how Instagram works, and why you see what you see' (Mosseri, 2021). These algorithm explainers may increase awareness of what algorithms do to sort content, but not necessarily increase control for ordinary users.

Clegg's piece has angered people for a number of reasons, not least because we know intuitively that Facebook has by far the greater impact on shaping the platform experience. Clegg falsely describes a scenario of individual control within what is actually a complex social and technical system. He ignores the fact that data and algorithmic literacies work in socio-technical contexts. This means that users can choose and negotiate data inputs to try to achieve certain algorithmic outcomes, but they do so experimentally in response to the parameters served up by the platform. The Facebook knowledge environment is one of great asymmetry. Users don't have full access to the accumulation and calculations made by the platform on the data that informs it.

In late 2021, nearly four years after Facebook introduced a new 'meaningful social interactions' (MSI) metric – with the advent of 'reactions' – new information circulated about how ranking and sorting calculations work. According to reports drawn from leaked documents, Facebook's response to falling user interaction was to re-calibrate its calculus of interaction types. It devised a new scale: 'each "like", for instance, would be worth 1 point; a reaction emoji or a reshare of a post without adding any text would be worth 5 points; an RSVP for an event would be worth 15 points; and comments, messages or reshares deemed "significant" would be worth 30 points' (Metz, 2021). Users can at best intuit the data inputs and signals. Facebook, on the other hand, gets to define, change, manipulate and operationalise its calculations. But likewise, the platform relies on inputs it cannot fully predict. Wherever this kind of asymmetry happens in digital environments there are likely to be associated data literacy gaps, and this is important for how we might actually 'learn to train the algorithm'.

Facebook and Instagram, along with TikTok, YouTube and others, have begun to offer short expository articles or videos ('explainers') in response to both the mounting public criticism of this asymmetry and the proliferation of algorithm 'folk theories' that have emerged to fill the gap. But such explainers are always limited in technical detail – platforms justify the vagueness on the basis of tensions between, on the one hand, the need to explain the algorithm in the interests of transparency and user literacy, and, on the other, the risk of inadvertently enabling bad actors to 'game' the platform. Petre and colleagues refer to this attitude as 'platform paternalism' (Petre, Duffy and Hund, 2019). A *Washington Post* exposé of Facebook's lean towards the negative explains this tension well:

> Facebook doesn't publish the values its algorithm puts on different kinds of engagement, let alone the more than 10,000 'signals' that it has said its software can take into account in predicting each post's likelihood of producing those forms of engagement. It often cites a fear of giving people with bad intentions a playbook to explain why it keeps the inner workings under wraps. (Merrill and Oremus, 2021)

This is not to say that users have no agency or control. Far from it. As Cotter has argued in relation to an analysis of the work that influencers do to negotiate Instagram's ranking algorithm, 'algorithms structure but do not unilaterally determine user behaviour' (2019, p. 895). Moreover, a large part of the mess Facebook has found itself in is due to the growing awareness of the ability of 'bad actors' to manipulate or exploit its algorithms. There are now so many competent operators on the platform that we should expect to see its algorithmic arrangements bend and buckle. However, the call to 'train your algorithm' as a frontline response assumes a kind of data access and expertise that is utterly unrealistic for the majority of any platform's user population.

Smaller, niche platforms are beginning to bring their target users along with them. Take, for example, Tryst (tryst.link). Tryst is a platform built to connect sex workers with clients – the sort of activity that Facebook's automated content moderation systems are now primed to shut down. Working more directly with, and for, sex workers, Tryst takes a more open and collaborative approach to explaining its search algorithm and making user data and analytics available and helpful for its users. It explains the parameters its dynamic algorithm uses to order search results from queries, ensuring that 'all providers at a given plan level get proportionate visibility, even in busy markets'.

> When someone performs a search on Tryst for a given area, we do a few things to determine what order they see the profiles in. After we filter for location, we then group profiles by their plan type [...] The profiles within these groups are then shuffled, presenting a different order to every user on every search. (Hunt, 2021)

The post goes on to detail how profile analytics works to help site users manage and improve their visibility. Algorithmic systems can be competitive and exploitative, or they can establish – as Tryst aims to do – a more transparent relationship based on subscription tier, profile information and location. But in general, ranking algorithms and recommender systems rely on a level of secrecy that is thought to be necessary to avoid platforms being gamed by users who understand and can manipulate the algorithm's weights and calculations. Everyday data literacies operate in this space of uncertainty between what we can and can't know, and how we choose to act in response.

Anthropological approaches to data and algorithms have helped to unpack the relationship and interactions between culture and technology, and how they shape each other. For Nick Seaver, rather than seeing algorithms as acting *on* culture it is more useful to approach algorithms *as* culture (2017, p. 4) – that is, they are artefacts of and agents in complex systems of meaning and value. And nor are algorithms defined purely within the technical domains that develop them: 'algorithms are not singular technical objects that enter into many different cultural interactions, but are rather unstable objects, culturally enacted by the practices people use to engage with them' (p. 5). This approach can be extended to the data that flows between people, platforms and devices and becomes the input and output for the algorithms powering society. Like other components of computational cultures, they are 'part of broad patterns of meaning and practice that can be engaged with empirically' (p. 1). This is where situated data literacies and cultural competencies play a key role in embedding these systems deeper into our everyday lives.

To get by, users of social media platforms have to gain *some* level of understanding and practical competence with the actions that the platform takes as signals, and the metrics and calculations that sort and rank content, and otherwise shape the platform experience. For Gibbs and colleagues, each platform has its own mix of 'styles, grammars, and logics', which they call 'platform vernaculars' (Gibbs et al., 2015, p. 257). But across social media platforms, there is increasingly little difference regarding what counts. As Twitter, Facebook, Instagram, Pinterest and Tumblr matured, their measures of activity and interactivity converged to align across particular kinds of data points (McCosker, 2017a). A famous instance was Twitter's move from the 'favourite' button (represented by a star icon) to a red, heart-shaped 'like', aligning the measure of engagement with a tweet with the protocols and metrics used by Facebook or Instagram (Newman, 2015). Understanding and gaining mastery over these protocols and algorithmic systems becomes part of the skillset of any aspiring YouTuber, TikTok star or Instagram influencer. But even for the rest of us, algorithm literacy is increasingly needed to operate as users, consumers and citizens in the contemporary digital media environment.

Consider Lil Nas X – a prominent example of achieving visibility within the contemporary music industry through a combination of vernacular data literacies. With success in the music industry now increasingly tied to algorithmic systems, artists who aspire to popularity arguably need to have an intimate knowledge of how platforms measure and shape visibility, as well as the skills to exploit these dynamics. In 2019 it was hard to avoid the endless looping of 'Old Town Road' on TikTok, YouTube and other platforms that build on the mass sharing of audio-visual content. Famously, Lil Nas X was seen by some as an 'imposter' who (inauthentically) 'gamed' social media in order to turn his song, 'Old Town Road', into a #1 hit, and winning a Grammy Award that year for Best Pop Duo/Group Performance, with Billy Ray Cyrus, and Best Music Video. Cultural writer Laja, who goes by @NotLaja on Twitter, investigated the way Lil Nas X worked with the measures and logics of various platforms' ranking systems. According to Laja's 2019 popular Twitter thread,[17] Lil Nas X achieved notoriety via savvy use of a range of visibility strategies, both old and new.

While in this example we're talking about celebrity culture, these are the outcomes of a highly skilled use and manipulation of data and algorithmic systems of communication and exposure. Even if Lil Nas X's tactics don't involve 'reading the data' directly, there are techniques at play that involve data and algorithmic awareness and manipulation of the metrics – making the moves that count. Lil Nas X placed the lyrics to 'Old Town Road' in various seemingly random places on the internet (including Reddit forums). This highlights the way content operates as 'data' on social media platforms. These 'product placements' were often in the form of questions about the lyrics that asked what song they were from, thereby generating inquisitiveness and hype around 'Old Town Road'. They also served to multiply the available data points for trend analytics, as well as search and recommendation algorithms.

There is a richness to the role that textual and visual data can play on platforms that goes beyond traditional marketing strategies. Just look at the way Pinterest groups and gathers content through its image recognition engines. Lil Nas X engaged in what @NotLaja calls 'image familiarity through artwork'. That is, he created consistently engaging and often witty imagery for each of his songs (featuring famous public figures, like Trump, Kim Jong-un and so on) that worked to tie this imagery to him in the mind of his viewers and followers. For @NotLaja, this was connected to his mastery of memes and natural gift for storytelling via Twitter and other forms of social media.

Algorithm literacy also involves working with or against the biases embedded in data-driven systems. Lil Nas X actively targeted white consumers of rap on YouTube – a demographic seen as key to signal boosting rap songs and increasing exposure for emerging rap artists. He was also seen to be clearly in tune with emerging cultural trends – such as the 'Yeehaw Agenda' in the case of 'Old Town Road' (Chow, 2019; Spanos, 2019) – an awareness that provided a focus for the most appropriate song of his to promote at any given time. He also performed a delicate balancing act around speaking to, capitalising on and keeping an ironic distance from his existing followers. Lil Nas X enjoyed a large existing follower base as a result of past activities as a Tweetdecker (Reinstein, 2018), and as the one (allegedly) responsible for a Nicki Minaj 'stan'[18] account. Stan accounts are accounts devoted to a pop star or celebrity. One of the noteworthy things about stan accounts is that, in addition to other more trolling behaviours, they actively work to 'stream-boost' short songs of approximately two minutes in length; perhaps not coincidentally, 'Old Town Road' is only 1:55 mins long. A combination of canny vernacular data literacy practices, then, led to 'Old Town Road' going viral.

Conclusion

In this chapter we have drawn together some of the key sites of everyday life – work, personal health and well-being, social media and popular culture – where we can see practices of data literacy at work in diverse, ordinary lives. Performance metrics and data-intensive forms of labour are affecting our relationships with institutions of work, well-being and sociability; there is increasing workplace and educational pressure to develop instrumental data literacies in response.

However, while there is great need to scrutinise the impact of these systems, and the very serious inequalities and injustices that are imposed by the platform economy, this is not the whole story. Since these data-driven systems generate responsive feedback loops by design, they also invite user involvement and result in new data competencies and vernacular expertise. Building on the work of cultural theorists and researchers who have provided accounts of situated literacies, we have shown that even everyday or vernacular literacies are not a magic solution to the social impacts of datafication and algorithmic logics, but they are resources for living with them.

Everyday data literacies can be understood as an emerging set of practices and data cultures that can contribute to oversight, self-protection and agency within disciplining, even oppressive, systems. But equally, everyday data literacies create new opportunities for social learning and collective activism, as well as for the amplification of malicious intent and action, as we will see in some of the examples of 'everyday data publics' in the following chapter.

5

Everyday Data Publics

In an era where social media has ushered in the unprecedented convergence (Burgess, 2017b) or 'context collapse' (Marwick and boyd, 2011) of our social worlds, everyday data cultures and practices play out not only in the private lives of individuals, but also in communities and in public. Just as in previous eras, publics are multiple and constituted via media and data as much as they are by formal governance practices. This is the key difference between contemporary ideas of issue- or interest-based publics and the idea of a relatively static (and not necessarily inclusive) 'public sphere', or the corresponding notion (beloved by national broadcasters and governments) of 'the public' (as in, the total pool of constituents for an organisation at the level of the state). When we come to see ourselves and act collectively through data, or in relation to data, then we can talk about being part of *data publics*.

Before aggregating up to the level of 'publics', however, we need to consider the range and meanings of terms that can describe the groups of people who generate data, or participate in everyday data practices (from counting their steps to attempting to manipulate a TikTok algorithm). Internet researchers (ourselves included) often refer to these people as 'users' – of infrastructures (such as Wi-Fi), or of specific apps, platforms and devices. The term has been handy for discussing contentious digital practices (such as the use of hook-up apps) because it doesn't attribute a sense of value or motivation. Additionally, it doesn't suggest a commercial relationship – that is, it lacks the implied dependency on markets and marketplaces which we might associate with the term 'consumer'. But while 'user' has its utility, it is not especially descriptive of the relational aspects of data cultures.

The term 'community' is often used by platforms themselves to refer to collective groups of users. However, these groups are rarely described in terms of their agency, or their capacity for collective action in relation to a platform's own policy statements. As Tarleton Gillespie notes, through their community guidelines (rules of participation), some platforms address their collective user population in libertarian terms, invoking principles such as freedom of speech and freedom of expression, and making it clear that the 'community' exists only within the container provided by the platform. In this context, community policies tend to imply that restrictions on content will only ever be imposed sparingly (and infrequently) (Gillespie, 2018, p. 48). More commonly, however, Gillespie observes that policy statements frame the platform itself as that which brings 'a diverse but fragile community' into being (p. 49). Where community is not recognised as existing outside of a platform, but is brought into being by the platform and then 'must be guarded so as to survive' (p. 49), there is a high likelihood that the 'community standards' will be those of the platform itself.

This framing of community differs somewhat from classic legal understandings of community standards expressed within policy and legislation – for example in obscenity law where 'community standards' dictate whether or not specific texts can be publicly circulated or sold. In these instances, both legislative documents and common law cases have typically included references to the abstracted (although tacitly raced, classed and gendered) figure of 'the reasonable person', whose values and perspectives are perceived to represent those of a broader population beyond the courtroom (Moran, 2003).

The fields of media and cultural studies have long been concerned with the ways collective groups might come to understand themselves in active or agentic ways through their shared experiences of media technologies – as audiences, as publics or both. As Sonia Livingstone explains, in common English usage, a public is constituted by 'a common understanding of the world, a shared identity, a claim to inclusiveness, a consensus regarding the collective interest', plus 'a collective and open forum of some kind in which the population participates' (2005, p. 9). As she notes, the body of qualitative research that emerged from the Birmingham Centre for Contemporary Cultural Studies in the 1970s found that media audiences engaged with popular news and entertainment content in ways that align closely with both common-sense and theoretical definitions of publics and publicness (Livingstone, 2019).

Further, while terms like 'public' and 'citizen' have historically been associated with nation-states, they take on new meanings in the context of globalised platforms such as Apple, Google, Microsoft, Amazon,

Tencent and Facebook that seem to be both everywhere and nowhere at the same time. They intersect with other categories (such as 'witness' or 'mob') that carry their own baggage in relation to historical hierarchies that promote some forms of collective action as civic dialogues, while others are dismissed as slacktivism or pile-ons. As far back as the 1920s and '30s, John Dewey was seeking a pragmatic approach to similar questions regarding human association and the pluralistic space of 'the public'. For Dewey, 'the machine age has so enormously expanded, multiplied, intensified and complicated the scope of the indirect consequences [of human association] that the resultant public cannot identify itself' (2016 [1927], p. 157). In other words, publics are not pre-existing or stable, but dynamic and emergent – and so it is with data publics (McCosker and Graham, 2018).

In their mapping of scholarly interest in the relationship between data and publics, Jannie Møller Hartley and colleagues identify four dominant approaches: a focus on the impact of data and datafication on the public sphere (as a site where collective understandings of citizenship and civic agency are shaped); an 'issues'-based approach that looks at the ways digital publics form in reaction to specific events or concerns; an interest in publics as sites of networked interaction that are shaped by platform affordances; and an examination of 'algorithmic publics' that are produced by platform sorting and recommendation systems (such as Facebook's 'People you might know') (2021, pp. 4–8).

It is clear, then, that the internet has provoked a rethinking of both publics and publicness – from danah boyd's (2010) conceptual model of 'networked publics' as a way to understand the early role of social networking sites in constituting collectives, to descriptions of the ways publics are constituted and engage as publics via hashtags, as for instance in Zizi Papacharissi's (2015) 'affective publics' or Bruns and Burgess's (2015) 'ad hoc publics', which, with the advent of algorithmic search and feed curation, become at least partly 'calculated publics'. One of the many examples of such publics relevant to the politics of everyday data cultures is #datamustfall, which, as Moyo and Munoriyarwa explain, 'marked the beginning of concerted citizen action against extortionist data pricing in South Africa' (2021, p. 375). Latterly, in relation to the intensively reflexive and performative dynamics of online publics in the context of algorithmic culture and microcelebrity, Crystal Abidin (2021) proposes the term 'refracted publics', where social activity is dispersed across different platforms.

There are certainly elements of all these conceptual models in the idea of 'data publics', but to begin with we can understand data publics more straightforwardly in terms of how data represents us and invites us to see ourselves as part of (or excluded from) various kinds of publics. These representations of us as publics *through* data also represent us in relation to the institutions and organisations providing and presenting the data.

In the first months of the global COVID-19 pandemic, as epidemiological information slowly emerged, public debate and contestation around the role of data was widespread in both legacy news outlets and social media channels. Daily updates on new diagnoses, hospitalisations and deaths were reproduced and circulated in public and private forums, as public health authorities sought to explain policy recommendations (from hand-sanitising and mask-wearing to lockdowns) with reference to statistical terminology and data visualisations. Twitter threads and posts on blogging platforms such as *Medium* began to emerge, challenging official predictions and promoting 'alternative' epidemiological modellings (Muir, 2020). Some of these were authored by experienced data scientists (without epidemiological experience), but others were DIY attempts by people with a basic knowledge of statistics (Muir, 2020). These forays into what might be termed 'armchair epidemiology' were strongly condemned by professional epidemiologists and mathematicians, as being misguided at best and 'carpet-bagging' (or self-promoting opportunism) at worst (Obungu, 2020). While DIY data science became less mainstream as the pandemic progressed, continuing debates around the optimal response to COVID-19 have revealed the extent to which data is both central to, and contested within, broader assumptions about public interest and 'the public good'.

These debates exemplify the tension between what Kennedy and Moss (2015) call 'known and knowing publics'. The contrast here is between the use of data and analytics to identify, measure and describe populations and publics on the one hand, and, on the other, what people might themselves come to know by understanding their collective situation through data. Everyday data publics constitute the wider circles of data practices, moving outwards from the intimacies and personal data literacies we have covered in previous chapters – although there are important crossovers. Epidemiologists, politicians and public servants use data to make recommendations based on their knowledge of demographics, geography, contacts between known positive cases, and biomedical data regarding viral transmission. That is, they seek to make sense of 'COVID publics'. Similarly, DIY data scientists and other public commentators seek to exercise agency and control, or demonstrate non-elite forms of expertise, through their own engagement with the data – however partial or incomplete it may be. This example highlights the struggles between expertise based in data mastery and the will to counter that expertise with vernacular knowledges grounded in everyday experience, and the ways in which these struggles can be amplified when personal data practices articulate to publics.

Data rituals

Even the most mundane personal data practices can connect us to public life, because they form part of what we might understand as communicative or media rituals. Writing in the 1980s, James Carey reminded his readers that this idea of communication as ritual had been part of (American) culture since the nineteenth century, alongside the more dominant idea of information transmission, which was fundamental to the mass communication paradigm (2009 [1989], p. 12). Similarly, we can understand data not only as information to be transmitted, understood and used to make decisions, but also as material for communicative and social rituals – the patterned and repeated practices through which social and cultural structures and meanings are embedded, maintained and contested.

Everyday data rituals, then, are mundane: patterned, repeated practices through which we structure our days and maintain our relationships. Back in the early 2010s, when interest in 'checking-in' to venues via mobile location-based social networking services was at its peak, a large part of the appeal of these services was that they articulated 'collective attachments', and enabled 'documentation of our relationship with the places with which we have special ties' (Schwartz, 2015, p. 96). In some cases, these place-ties were reinforced through highly ritualised yet routinised day-to-day interactions, such as walking a specific route in order to check-in at a favourite public building, or queuing up an app before leaving home in order to check-in while driving over the Golden Gate bridge (Wilken, 2019, p. 141).

Building on the work of the anthropologist Catherine Bell, Lee Humphreys (2018) draws on a range of examples from blogging to diaries to demonstrate that everyday practices of media documentation – which she calls media accounts – can turn mundane practices into media rituals (Humphreys, 2018, p. 35). Humphreys also draws on Karin Becker's (1995) work on everyday media rituals to note that in social media environments, mediatisation can directly transform such private rituals into public ones. In this way, everyday digital media rituals link individual or apparently private experiences (such as those taking place within a single household) to public life. Recent years provide examples of convergent digital media rituals that bring together everyday media consumption, public media spectacles and hashtag publics – from the Eurovision song contest (Highfield, 2017), which sees in-person watch parties remediated via Instagram selfies and shared on Twitter using the #eurovision hashtag, to the public outpouring of grief articulated to personal life stories that surrounds the deaths of much-loved celebrities like David Bowie (Burgess, Mitchell and Münch, 2019).

The annual Spotify Wrapped event that we discussed in Chapter 3 articulates mundane, everyday media rituals to 'calculated publics' by nudging users (via buttons that result in automatically formatted posts) not only to explore their annual analytics through visual dashboards, but to post their top results to popular social media platforms. The event and its accompanying data visualisations produce much more than a pleasurable interactive musical portrait of each user's listening habits over the past year: the connective dynamics of its in-built hashtags (#spotifywrapped and its variations, like #2020wrapped, used in December 2020) convert it into a manufactured digital media ritual with data at its heart.

Analogously to Dayan and Katz's famous (1994) description of major public events like inaugurations or royal weddings as broadcast media's 'high holidays', Spotify Wrapped is an attempt to create a 'high holiday' in the platformised media environment. It explicitly invites (and actively nudges) individual consumers to participate in social media publics on the basis of their shared but diverse experiences of engaging with the platform's 'data selfies' dashboards and analytics, generating a secondary aura of mainstream media coverage (Galant, 2020; Weiss, 2018) for the platform. It is a prime example of the shift from *ad hoc* (apparently organic, emergent) hashtag publics to *calculated* ones (Bruns and Burgess, 2015) – that is, the platform nudges us to use its temporary affordance to post publicly about our unique results in a generic format shared by millions of others; through our collective participation on the hashtag we make the leap from deeply personalised data intimacy to membership, however fleeting, of a data public.

The spectacular and synchronous character of Spotify Wrapped as a media event also prompts social learning (discussed in detail later in this chapter), as users and journalists try to understand the algorithmic recipe for the individual dashboards each user sees, and what the company might be doing with all these analytics on the back end – a learning process that Spotify and its internal experts actively promote and participate in. For example, the marketing-oriented data science publication *Springboard* marked the launch of the 2020 edition of Spotify Wrapped with an article that aimed to 'break down how Spotify Wrapped works, why it's so effective, and how Spotify is leading the charge in *creating an emotional connection with its consumers through data*' (Galant, 2020, our emphasis). Walking through the 'strong' emotional design and dynamic interactivity of the dashboards, and the campaign's simple, automated social media sharing affordances, the article then turns to the data operations that underpin not only the Spotify Wrapped event, but Spotify's personalised recommender systems more broadly, along the way providing explainers of key data science techniques: collaborative filtering, Natural Language Processing (NLP) and AI-enabled audio analysis.

As these brief examples show, digital media rituals are both routine and dramatised ways of working through, performing, distancing ourselves from and testing the new social arrangements associated with digital trans-formation. In this sense, then, when we get involved in digital media rituals and become part of data publics, we are taking part in the collective effort to figure out how to live together with data and machines.

Data activism

Platforms treat user-contributed content as data to be measured and judged by automated systems. In doing so, that content is weighed against concerns that shape the 'standards' – both legal and normative – governing what is shown and to whom on the platform. Despite the dispensation platforms are given as 'mere hosts' of content, their established community standards hit on socially debated issues concerning the treatment of race and gender, political stance, mis- dis- and mal-information and personal well-being. In this way, social media platforms have become lightning rods for public contest over social divides and deep-seated social problems. And it often takes the collective effort of users banding together to observe and challenge the social impact of a platform's algorithmic treatment of content and seek to affect change.

In April 2019, Instagram confirmed that it was deploying a policy of 'demoting' posts that were deemed 'sexually suggestive' but did not explicitly violate its community guidelines (Constine, 2019). The announcement was made at a Facebook event, with Instagram staffer Will Ruben announcing, 'We've started to use machine learning to determine if the actual post is eligible to be recommended to our community' (Constine, 2019). As Josh Constine of *TechCrunch* reported, Instagram creators were dissatisfied with the vagueness of this policy, which they experienced as censorship and suppression. These Instagrammers were diverse – some were sex workers using selfies to direct new audiences to their more explicit accounts (such as OnlyFans). Others were fitness instructors and performers posting images of themselves in dance costumes and workout clothes. Visual artists and photographers whose work included non-explicit nudes (often with an overt agenda of body positivity) also experienced drops in engagement – as did a range of birth and parenting-focused accounts, including doulas and breast-feeding consultants.

While these creators were (in some cases) targeting quite different audiences, their content was treated in a similar way by Instagram's content moderation practices, which appeared to uniformly identify all naked or semi-naked images as 'sexually suggestive'. Creators of 'suggestive' content complained that even when they actively worked to understand and abide by the community guidelines, they were 'shadowbanned' – that is, their content was hidden from search results and recommendations (including Instagram's Explore page). Throughout 2019, members of this nascent public talked with each other and with their followers about shadowbanning. Dialogues took place on blogs and comment threads, with Instagrammers sharing their concern that content was being hidden, based on their observation of variations in engagement numbers on differing posts.

Activists who made direct inquiries to Instagram regarding which poses or hashtags might be affected were initially informed that their content was not, in fact, being demoted (Are, 2019). During the same period (mid-2019), concerns grew among Black, queer and gender-diverse Instagrammers regarding what appeared to be a policy of automated content moderation that was leading to their content being incorrectly flagged as 'explicit'. For example, an application for paid advertising to promote the queer and feminist magazine *Salty* (featuring an image of a fully clothed Black trans model) was refused by Instagram in a message citing policies regarding 'escort services'.[19] Other LGBTQ+ companies were similarly refused advertising on the basis of 'adult content' during this period (O'Hara, 2019). Digital sex-work activists organising in the United States under the banner of 'Hacking//Hustling' linked increases in shadowbanning to the 2018 implementation of FOSTA-SESTA, a legislative change to the US Communications Decency Act which made platforms legally liable for content related to 'sexual solicitation' (Blunt and Wolf, 2020; Blunt et al., 2020).

Throughout 2019, content producers increasingly worked together to circulate information and 'test' suspected shadowbans by searching for each other's accounts and collating the results of those searches. They similarly invited followers to assist by searching particular posts and hashtags and reporting back on whether 'problematic' posts were visible in feeds. A team of Instagrammers and activists including sociologist and pole dancer Carolina Are (2020, 2021) founded the EveryBODYVisible campaign, inviting content providers and their supporters to post protest content, with the claim 'our movement includes artists, photographers, athletes, dancers, yogis, entertainers, strippers, sex workers, the LGBTQIA community, the BBW community, feminists, disability activists, body-positive, sex and birth educators, fitness professionals and many more'.[20]

The activist group produced a series of original semi-naked images and multiple versions of the EveryBODYVisible logo, which could be overlaid on supporters' Instagram posts, and shared a call to action for an organised protest, inviting supporters to post content (tagging Facebook and Instagram staff) on International Internet Day (29 October) 2019. Campaign images were widely circulated by high-profile Instagram users including burlesque dancer Dita Von Teese, but were (in some cases) themselves subject to moderation and takedowns on Instagram. Documenting the campaign in her blog, Are described the diverse strategies deployed by EveryBODYVisible:

From seeing women's engagement increase when they changed gender to male on the platform, to tagging the chiefs of Facebook and Instagram during our campaign, so that their 'tagged' section on their profile ended up featuring a variety of women's bottoms, we made so much noise that Instagram CEO Adam Mosseri had to recognise our demands were reasonable in a story post. (Are, 2020, p. 742)

Similar issues and collective responses have emerged in relation to the perceived treatment of body size through Instagram's AI moderation (Richman, 2019). Throughout 2019, the founder of The Visible Collective, Jessica Richman, fielded numerous accounts of Instagram's unintentional discrimination against fat bodies. Speculation circulated that Instagram's automated systems were calculating ratios of exposed flesh and identifying body parts such as 'female-presenting' breasts or nipples. While Instagram's parent company Facebook denied that its automated systems calculate the percentage of flesh – and hence disproportionately affect women showing larger bodies – it was clear that moderation policies regarding 'adult content'[21] were at odds with body-positive uses of the platform. Plus-size model and performer Carina Shero (currently @unskinnyhero on Instagram) described having multiple accounts flagged and banned by Instagram and having to work the system carefully (Richman, 2019).

The collective responses can be understood as examples of everyday data publics in action. The Adult Performers Actors Guild (APAG) set up a page to help those who feel they have been 'wrongfully deleted ... fight back against the unfair and discriminatory tactics'.[22] By collecting experiences and banned content, APAG seeks to establish a dataset to counter and redress platform moderation practices that unfairly target their community. In 2020, Instagram responded to plus-size model Nyome Nicholas-Williams, photographer Alex Cameron and campaigner Gina Martin's push for its policy to be reviewed, with the platform agreeing to allow women to present their bodies naked while cupping or hugging their breasts (a pose that had previously been flagged as pornographic).

Collective data action and strategies of circumvention can often be seen on platforms as workarounds to the way they moderate various kinds of 'borderline content'. This is the kind of content that platforms consider problematic but not within their scope for banning outright. They might at any time and without public acknowledgement restrict the visibility of hashtags or keywords that signal certain types of content deemed undesirable, by hiding or throttling their visibility in search and feeds. Content suppression, as Tarleton Gillespie argues, is a fundamental and in some ways ordinary part of the work platforms and their recommender and timeline algorithms do in ordering and organising content.[23] These practices differ from shadowbanning because of their focus on types of content shared, rather than on accounts or profiles targeted as breaching standards or rules. It's difficult to see or predict which content a platform will downplay or promote through its sorting models – which themselves change according to policy, politics and events.

In addition to sexually suggestive content, consider the treatment of topics concerning mental illness, which has been treated by Instagram as tending to involve problematic or 'borderline' content by default (McCosker and Gerrard, 2021). The hashtag #depressed and others on the same topic have been periodically banned or searches on these terms suppressed, and related mental health tags showing cutting and other forms of self-harm, or content referring directly to eating disorders or '#thinspo' (used to mark content providing 'inspiration' for weight loss, thereby promoting disordered eating), are consistently hidden or removed. Content-curation strategies in these instances are worthy attempts to improve well-being through platform use. However, empirical work shows that not only is #depressed used as a collective touchpoint for huge numbers of most likely young Instagram users, those users can be very creative in shaping posts to both signal depression – albeit anonymously – and play on sad sentiment or subvert it through sad and 'dank' memes (popular in-jokes playing on over-used memes and passing trends) for 'relatable laughs' (McCosker and Gerrard, 2021). By collectively finding the means to circumvent the markers that lead to mental health content being removed or hidden, adaptive cultures of platform use thus create new ways of connecting with and through that content.

Collective data practices and responsive algorithmic practices like this can also take shape through memetic logics and genres of communication. Memes become the mechanism used to circumvent platform moder-ation triggers, while fostering knowing publics through their multiple and culturally rich modes of meaning-making. #depressed memes on Instagram are common. Some researchers have considered memes and 'dark humour' in relation to #depressed content as the 'noise' of platform use (Ging and Garvey, 2018). Others have shown the way memes serve to generate bonds around mental health content through the collective manipulation of the signals that trigger, or circumvent, algorithmic moderation (McCosker and Gerrard, 2021). Following Beraldo and Milan's notion of the 'contentious politics of data' (2019, p. 3), we could say that these are illustrative of the 'everyday practices of resistance, subversion and creative appropriation embodied by individuals' in their response to platform restrictions.

Similar public oversight of platform metrics, automation and curation has targeted Twitter's data operations. Twitter users began to notice and then to document the apparent racial and gendered bias built into the machine vision system that the platform used to crop faces in images posted in tweets. Large numbers of Twitter users began publicly experimenting with images to show themselves cropped out in favour of, for instance, attractive white men who also happened to be in the image. PhD student Colin Malard discovered the bias when he took to Twitter to complain about Zoom's 'virtual background' feature failing to recognise his Black colleague and visually erasing him from a conference call (Hearn, 2020). When Malard posted the screenshot to Twitter, it also cropped his colleague out, sparking a sustained trending conversation and subsequent public experiments with the automated cropping system. Twitter user Tony Arcieri, for example, posted an image of Barack Obama and Mitch McConnell, with the caption 'Trying a horrible experiment ... Which will the Twitter algorithm pick: Mitch McConnell or Barack Obama?' (Hearn, 2020).

As Sarah Williams explains in her book *Data Action* (2020), participation in data-oriented action, as in the examples above, also has a way of producing publics or strengthening communities and affected groups. By working with data to defend shared interests, data publics emerge and evolve rapidly, their impact sometimes forcing changes to the technologies shaping their everyday experience. Faced with the extensive negative publicity in relation to its image-cropping algorithm's racial bias, Twitter was relatively quick to apologise, admitting that it had work to do on its automated systems, and offering incentives for public contributions to this task. With all the internal testing and analysis at its disposal, it was the public surfacing of data-derived inequalities that mattered.

The intensity of the responses to Twitter's sorting of visual data, and to the treatment of users on Instagram, demonstrates both the extent to which those inequalities are felt as problems among affected communities, and the affective capacities available for counter-publics to push back. However, with the rapid development of data-hungry technologies and tools, it's not all smooth going resulting in productive pushback against harmful platform policies. As we learn to live with data collectively, the potential for malicious and disruptive data-driven applications also increases.

Social learning

Social media platforms have created the conditions for observing new forms of social learning and extending them to generate new publics and counter-publics. Data and its manipulation are the foundation, and often feature as a source of collective struggle, as the examples above show. With those responses to the way platforms shape social action and experiences we can see how social and computational learning are always interacting. Fourcade and Johns point out that the interaction of social and machine learning is about more than the development of skills and knowledge. Learning is extended 'through imitation, stylistic borrowing, riffing, meme-making, sampling, acculturation, identification, modeling, prioritization, valuation and the propagation and practice of informal pedagogies of many kinds' (Fourcade and Johns, 2020, p. 806). Social learning extends and generates new public formations, and also creates the conditions for potential manipulation – both productive and malicious.

These ways of knowing and acting are entangled with the technical systems that facilitate interaction and the metrics that determine how that interaction becomes meaningful in a different way, as data. Think about the speed at which trends emerge and dissipate through social media. Platforms like TikTok are predicated on offering the tools for propagating actions and reactions in memetic form as trends that take on a life of their own. Along the same lines, informal pedagogies have proliferated in YouTube 'how to' and tutorial genres. In addition, on Reddit and elsewhere there are threads dedicated to sharing knowledge on how to perform 'location spoofing'. This usually involves the use of VPNs, location scrambling apps and/or other means in order to mask one's location when using a smartphone to access specific apps. This informal knowledge can be put to use in a variety of ways, from the largely benign and mischievous (say, when using Pokémon GO in order to 'catch them all') to the more concerning (say, when using dating apps such as Tinder). In the previous chapter we considered these peer pedagogies through the lens of data and algorithmic literacies. Competencies and skills are developed through knowing data and the way content and digital interactions are organised as metrics, and the operations of algorithms across social media platforms. These competencies, skills and knowledges are personal and individual, but can also be rich in social context and collective sense-making. And through the rituals of digital interaction and data use we collectively take part in learning how to live with data and machines.

The concept of social learning has been used for many decades as a counter to individual cognitive and knowledge-based theories of learning, emphasising instead the importance of observation, imitation and modelling (Bandura, 1969; Vygotsky, 1980). The flow of data through our everyday use of digital technologies, combined with developments in machine learning and algorithm design, mean that interactions between social and computational learning are already extensive and complex. Drawing on half a century of learning theory, including the work of Vygotsky, Fourcade and Johns note that 'machines are poorly equipped to deal with the fact that all human learning is cultural, that is, anchored not in individual psyches but in collective systems of meaning and in sedimented relational histories' (2020, p. 807). Social learning takes many forms and happens now across a wide range of platforms and sites (including GitHub, YouTube, TikTok, Reddit), from formal to informal, through academic research but also through 'folk learning' (Pangrazio and Sefton-Green, 2020).

Consider the case of deepfakes: synthetic visual and audio media, generated through the acquisition of large datasets and deep learning systems designed to synthesise or create entirely new outputs. The term deepfake entered popular discourse in 2017 through the activities of Reddit user u/deepfakes, who also created the subsequently banned subreddit r/deepfakes. With source code available through open repositories – primarily the code-sharing platform GitHub – deepfake creation was possible for those with only moderate coding and data-management skills (Winter and Salter, 2020). Endless hours of online digital video offer source data ready to be broken down into tens of thousands of stills, cleaned and managed, making it possible to synthesise faces with alternative bodies, or reshape expressions, gender, age and any other personal identifier. With the distribution of code, data and instruction and advice through Reddit and other forums, and their early uses in synthetic porn on platforms like Pornhub, deepfakes have been embedded from the beginning in the 'macro context of gender inequality' (Öhman, 2020, p. 133; Winter and Salter, 2020).

Fuelling their problematic uses and risks, but also inspiring a range of counteractive techniques, deepfakes have become a key site for building coding literacy (Vee, 2017), socialising AI systems and learning about the manipulability of visual data. Learning to deepfake, extracting, working with and manipulating visual data, has become a way of learning collectively *with* machine learning models. As products of informal pedagogies and shared code repositories and image datasets, deepfakes are cultural artefacts. They play most heavily off cultural touchstones – like the distinctive mannerisms of the actor Tom Cruise, endlessly parodied on TikTok through the work of Belgian VFX specialist Chris Ume @deeptomcruise (Vincent, 2021a). Ume's Cruise deepfakes, made in collaboration with Tom Cruise impersonator Miles Fisher, are time-consuming creations requiring a high degree of expertise. However, with simple interface tools available, and an extensive YouTube 'how to deepfake' genre, these kinds of synthetic media are generating substantial flows of social media activity. This has played out in multiple directions – as entertainment, with malicious intent and harm (particularly in non-consensual image manipulation and sharing), as well as in computational attempts to detect and disrupt artificially altered video. In other words, these are data practices that have emerged through platform cultures underpinned by social coding, expanding our sense of everyday data and AI literacies into shared, social and collaborative domains (McCosker and Wilken, 2020).

Similarly, in the collective work of malicious actors and bot networks on platforms like Twitter and Reddit, we see collective learning in the manipulation of systems, signals and data as they trigger and respond to metrics and algorithmic sorting. The actions of coordinated bot attacks work to counteract moderation measures and mimic and subvert organic norms of information distribution. Through coordinated, disruptive activity, bot networks generate content and distribute or amplify controversial information in order to disrupt political systems and, for example, public health messaging (Broniatowski et al., 2018; Graham et al., 2020). The motivations behind these networks are often not known, and their activity usually involves a mix of malicious actors (actual people) and automated accounts created to look like those of real users.

Coordinated networks like these can distort the public formations that emerge in relation to important and sensitive issues. They operate on and affect information flows in ways that exceed the capacity of individuals – so they come to count through metrics like trending topics. This collective effort also generates a level of group visibility, which if sustained long enough can register further attention and attract potential amplification. And they spark highly affective, emotional activity for or against, with or in response to their provocations (McCosker and Graham, 2018; McCosker, 2014) – they might be considered to be more-than-human data publics in themselves.

Some automated accounts (bots) on platforms like Twitter are built to act as if they are human, posting often artificially constructed content, sharing links, liking other posts and re-sharing them, and commenting.

In this sense they activate and artificially manipulate all the signals the platform uses to measure activity and promote trends and trending information. Contra the dominant narrative, the conversion of platform use into metrics (achieved through the embedding and then datafication of platform affordances that were originally invented by users – such as hashtags and @ replies) invites and rewards these activities (Burgess and Baym, 2020, pp. 111–12). Accounting for data publics, then, must include the influence of these kinds of coordinated networks of actors.

Even before the global outbreak of COVID-19, coordinated networks of bots and people were deliberately aiming to disrupt the flow of public awareness and discussion of the benefits of vaccination (Broniatowski et al., 2018; Yuan, Schuchard and Crooks, 2019). As a movement cohering around belief in the harms of vaccination or in the personal right to refuse public health initiatives, antivaxxers act through coordinated data publics to counter vaccination messaging, using the same tools of hacking trends and visibility metrics as activists seeking to right the racial bias against or targeting of marginalised groups. In an analysis of more than 25 million tweets over a ten-day period during the COVID-19 outbreak, Graham and colleagues identified 5,752 accounts coordinating the spread of mis- and disinformation 'for either commercial or political purposes' (Graham et al., 2020). They attribute this activity mainly to the promotion of right-wing political parties or governments. As well as bringing into being potentially harmful counter-publics disrupting public health efforts to protect populations against the deadly coronavirus, bot networks, like deepfakes, have created communities of researchers, amateur sleuths and hackers creating detection tools to expose and respond to them.

During the pandemic, antivaccination campaigns also became entangled with opposition to 5G technologies (Bruns, Harrington and Hurcombe, 2020). This has taken several forms. There have been coordinated misinformation and disinformation efforts, such as circulating a conspiracy theory suggesting that 5G network radiation directly causes COVID-19 (Meese, Frith and Wilken, 2020). There have been public protests to block or undermine the construction of 5G-related infrastructure (in the UK, these protests took the form of direct action, with at least seventy-seven cell towers set on fire between early April and early May 2020 [Asher Hamilton, 2020]). And there have been coordinated efforts directed at policy-makers and governments, attempting to delay or halt 5G implementation, mostly on health grounds (Meese, Frith and Wilken, 2020). These efforts have had real-world impacts. In Australia, both federal and state public health officials have had to issue public statements aimed at combatting misinformation around the causes of COVID-19 (Meese, Frith and Wilken, 2020). In Europe, some countries have paused the roll-out of 5G infrastructure in response to public health concern, with Switzerland announcing an 'indefinite moratorium' on the use of 5G technology (Jones, 2020).

Developing the idea of a 'contentious politics of data', Beraldo and Milan (2019) draw attention to the ground-up transformative initiatives that can interfere with or hijack dominant processes of datafication. Aligning with Dewey's notion of emergent publics (rather than 'the public' as pre-existing), their work is a reminder that the politics of data can be enacted by individuals or collectives, but that collective action is never a given, and is always something that has to be worked towards. Similarly, activism shapes data, and data and platforms shape activism. In other words, data can be the object of intervention and give rise to the need for collective action, or it can be the instrumental or strategic means for undertaking collective action or advocacy. In the first instance, data is what's at stake; in the second, data is the medium and repertoire for collective action (Beraldo and Milan, 2019, p. 6). So, on the one hand, as Flyverbom and Murray argue, 'largely invisible and seemingly technical ways of dealing with data have implications for how individuals, organizations and societies get access to worlds and realities, curate their presence, and carry out interventions' (2018, p. 2); while, on the other hand, we see renewed and broadening attention to what is at stake when data is the medium through which everyday life is carried out.

Public art and data visualisation

Data is the focus of some public art practice, where the aim has been to generate greater public awareness of datafication. Similarly, creative or interactive data visualisations are used by a range of organisations, and increasingly in 'data journalism', in order not only to generate knowledge about publics but to promote '*knowing* publics' (Kennedy and Moss, 2015, emphasis added). However, as the projects discussed below show, there is a persistent tension in data cultures between formal expertise (of the data scientist or artist) and everyday experience (of the data subject), one which remains unresolved.

In his book *Living in Data* (2021), data visualisation artist Jer Thorp explains his evolving goals across nearly two decades of working with communities and data visualisation. Like others working with social data at the community level, Thorp argues that social data analysis should be 'less about finding answers in data and more about finding agency'. For Thorp, data is always tangled with the quotidian, affecting how people work and live. He describes the practice of working *with* communities, rather than just representing them through data analysis and visualisation, and sees in this the opportunity for 'question farming', where data visualisation is used not to simplify something in the world, but to uncover complexities and expose social phenomena previously unseen and unknown (Thorp, 2021). Data visualisation, however, has always carried the pretence of objectivity, and professed a taken-for-granted explanatory power garnered through conventions of statistical communication (Kennedy et al., 2016), even when it involves contentious data politics and complex or opaque forms of analysis.

The social role of diagrams, graphs, maps and charts of data visualisation has traditionally been to make data accessible to 'the public'. Increasingly, however, with the availability of large datasets and tools for analysis and visualisation, along with new professions like data journalism, 'their novel forms and uses mean that our understanding of how they work as semiotic and aesthetic phenomena and how they support or hinder personal and social agency is also in flux' (Kennedy and Engebretsen, 2020, p. 20). Data visualisation comes in many shapes and forms, but with computational tools and vector graphics it has become precise and even aesthetically beautiful (Halpern, 2015). Websites like Flowing Data,[24] or the subreddit r/DataIsBeautiful celebrate the aesthetic techniques of contemporary data visualisation, potentially cultivating public engagement.

In fact, public data visualisation might be increasingly part of most people's everyday media experience – as with the graphs and charts tracking COVID-19 cases or vaccination rates for example – but this does not necessarily displace the complexity of big data or overcome the potential for public misuse or misinterpretation. There is still a tension between the legibility of data visualisation and its role as a stand-in for statistical and mathematical operations, or data calculations, that exceed 'ordinary' comprehension. Big data in visual form operates, that is, between the sublime and the diagrammatic – between the reduction of data to the symbolic, and its materialisation as something that can be used to ask questions or solve problems (McCosker and Wilken, 2014). Indeed, the 'feeling of numbers' is vital to the way publics engage with data (Kennedy and Hill, 2017), and to their power to act.

By devising similar methods and tactics in collaboration with civil society organisations, journalists and citizens, research projects and centres can work to bridge the gap between the opacity of everyday data and the subjects of that data. In their Art/Data/Health project, Fotopoulou and colleagues have sought to 'build a bridge between data, creativity and experiential stories', bringing communities, artists and civil society organisations together to improve collective agency and ownership of health and well-being data.[25] In the Living With Data project based at the University of Sheffield, Kennedy and colleagues draw on the 'data journeys' developed by Jo Bates to make visible particular data practices at moments in time, to help identify and explore everyday data and what people know and feel about it.[26]

The power relations inherent in the tension between 'known' and 'knowing' publics, though, are not a simple matter. For example, Dorothy Kidd (2019) explores data activism and data sovereignty through the counter-mapping practices of Indigenous nations in Canada. As a reminder that data activism remains embedded in a whole ecosystem of economic, social and cultural relations, and that counter-mapping as a form of data justice requires 'fusion within larger projects of redistributive, transformative and restorative justice', Kidd documents the way the Inuit, Dene, Wet'suwet'en and other First Nations people in Canada use 'mapping practices, together with narratives, songs and prayers, to re-inscribe historical memories and revive their long-standing reciprocal relations with their territory and ways of life' (2019, p. 966).

Conclusion

This chapter has addressed the changing ways in which we have come to know ourselves *as* publics, and *act* as publics, through data. We have argued that data practices are not only personal and individual but also, crucially, collective and social. Data is increasingly used and visualised to know groups, populations, users and communities. But it also helps to constitute 'knowing publics' (Kennedy and Moss, 2015). The cases covered in this chapter involve collective responses to the data that surrounds us, as well as the personal data practices, as in the case of Spotify Wrapped, that connect us to public life and online publics by fulfilling the role of important communicative media rituals.

Everyday data publics can be found in the collective efforts of users banding together to observe and challenge the social impacts of platforms' algorithmic treatment of 'contentious' content. We have detailed how media platforms create the conditions for observing and enacting new forms of social learning, and how these have been extended to generate new publics and counter-publics. Despite earlier enthusiasm about the role of networked publics in driving positive social change, these practices also empower groups with nefarious intentions. Whether for good or ill, collective action is facilitated by the way platforms gather together their activity data and interpret it in terms of trends. We have also seen the social role of data visualisation and public data art in constituting and fostering these known and knowing data publics. Through our engagement with this diverse material, we have sought to trace the data cultures that form and dissipate across social media platforms, and to demonstrate the power – positive and negative, productive and destructive – in the formation of everyday data publics.

6
Conclusion

Generations of scholars have traced the role of media and popular culture in everyday life through successive waves of technological change. We hope this book has shown that the relationship between data and everyday life in the twenty-first century is just as dynamic and historically consequential as was the role of media in the twentieth, entailing multiple sites of ongoing struggle, including over the extent and characteristics of human agency in relation to media and technology power. In this we agree with Livingstone (2019) that 'including the people in a mediated, perhaps mediatized, increasingly datafied age' – with all its dehumanising discourse – is 'the task in front of us'. With this book, we hope to have contributed something to the effort.

Throughout the book we have provided detailed accounts of the way datafication is ever-present in daily life. At the most mundane level, flows of data are generated and used in our daily lives as we move around, do our work as citizens and consumers, and relate to each other, in the most ordinary ways. As we have worked our way through the book's themes and examples, we have encountered the messiness of data in everyday life – the scraps of paper, inadequate programs, exploitable glitches and mismatched profiles that mean proprietary promises of frictionless tracking, sorting and monetisation are never fully realised. We have seen how the data logics and algorithmic operations of the big digital platforms are sometimes sites of struggle and negotiation. In everyday life, there is no sublime space of pure technology – there are only everyday data cultures.

As we saw in our discussion of sextech and smart homes in Chapter 3, data is entangled with the intimate relationships we have with ourselves and with each other, with the cultural products and practices that move us, make us think, or that simply, as Spotify would have it, 'soundtrack our lives'. In data's cultures of use we also find creative, playful and even joyful practices. These range from Max, a trans-masculine person in a small regional city, gaming Tinder's gender settings in order to find new queer friends and lovers safely (Chapter 3), to Lil Nas X's manipulation of social media's attention economy (Chapter 4), to the EveryBODYVisible coalition's day of action that filled Instagram executives' 'tagged' pages with pictures of naked butts (Chapter 5). Through these examples – and others – we have sought to demonstrate that data is a resource for learning and knowing publics (Kennedy and Moss, 2015) as much as it is a challenge to their well-being.

We also showed how, in personal, intimate and public digital media contexts, data can be used in abusive and discriminatory ways just as much by ordinary people as by large global corporations. We saw this in the discourse that promotes domestic spyware as 'care', and in our discussion of the lengths that Max and other dating app users have to go to in order to avoid being targeted and harassed by their peers on the basis of their perceived race, sexuality or gender.

Throughout the book, we showed how the politics and ethics of everyday data cultures can involve the users of technology as participants, sometimes unwittingly, sometimes reluctantly, and sometimes playfully, skilfully or purposefully. Building on these insights, we now turn our attention to the pragmatic questions entailed by our theoretical framework, and discuss how we might go about gleaning 'resources of hope' (Williams, 1989) from everyday data cultures.

Hopeful data futures

As we discussed in the Introduction, most critical data studies scholarship to date has been primarily oriented around diagnosing the social dynamics and dangers of datafication, and raising public awareness of, and encouraging resistance to, these dangers. The next steps – or the answers to that most venerable question, 'what is to be done?' – are harder to articulate, it seems. Indeed, as numerous reviewers of Zuboff's 'magisterial' (2019) work (as synthesised in Jansen and Pooley, 2021) agree, *The Age of Surveillance Capitalism*'s final 'call to action' chapters fall flat, and the conclusion is 'less a call to arms than a general wake-up call' (Varghese, 2019).

When pragmatic suggestions for action appear in the conclusions of these works at all, they usually fall into two categories: what is needed, we are told, is more literacy, more regulation, or both. The first of these usually argues not only for readers to continue to educate themselves, but also for more initiatives aimed at improving critical data literacy in the population (e.g. Kitchin, 2021). The second category calls for further activism aimed at effecting stronger state-based regulation. At the conclusion of *Atlas of AI*, for example, Kate Crawford calls for 'a renewed politics of refusal', and invites 'the next era of critique' to '[overturn] the dogma of inevitability' (a dogma which implicitly endorses the unstoppable expansion of AI) (2021, p. 226). While the discourse of resistance and refusal dominates Crawford's account, she also sees hope – there are, she writes, 'commons worth keeping, worlds beyond the market, and ways to live beyond discrimination and brutal modes of optimization' (p. 227). But, beyond 'refusal' and activism in support of greater regulation, we are left wondering what role ordinary people are meant to have in all this; and what pockets of hope and strength-based resources might already exist in their lives.

Another place to look for resources of hope is in the everyday data work conducted within and on behalf of the tech companies driving many of these developments, as well as within a range of other industry, government and community organisations. We might think of these forms of data work, and the meanings and values associated with them, as an element of data and AI's *production cultures* (see, for example, Stark and Hoffman, 2019), to adopt terminology from the field of media industry studies (Herbert, Lotz and Punathambekar, 2020). These production cultures are also the scenes of everyday life, and of everyday activism, for tech workers. Mary Gray and Siddharth Suri's book *Ghost Work* (2019) brings a powerful – and dignifying – labour critique to bear on the manual tasks of data wrangling and processing that are often hidden beneath the mythology of autonomous AI, whether in the form of Amazon's Mechanical Turk workers (Turkers) undertaking data labelling tasks, or content moderation workers subcontracted to Google or Facebook (Roberts, 2019).

The public controversy around the firing of Timnit Gebru from Google's AI ethics team (Hao, 2020) provoked a wider reckoning around 'ethics-washing' (cynically adopted and superficially implemented ethics principles) in tech companies. The stories of the circumstances that led up to Gebru being fired also revealed signs of internal perturbations of the dominant techno-culture within Google (Hao, 2021; Simonite, 2021), at least some of which relate to the 'everyday activism' (Vivienne, 2016) of Googlers based on their experiences as racialised and/or gender diverse people. So far, however, these do seem to be perturbations rather than seismic shifts – suggesting that the profit-and-growth imperatives of techno-capitalism still overdetermine their data cultures in the end. While the detailed empirical study of these production cultures is beyond the scope of the present book, more research that seeks to understand and enhance the relationships between the lived experience of data workers – their 'whole ways of life', and not only the conditions of oppression under which they labour – and the shaping of the automated decision-making systems their labours support is much needed.

But the friction and increasing public visibility of these controversies also provide some cause for optimism. For example, trans researchers are leading the development of more nuanced technologies for gender and voice recognition, or what Alex Ahmed (2018) has termed 'trans competent interaction design'. Oliver Haimson and colleagues (2020, 2021) have worked with trans/non-binary people to scope the potential for technologies that 'allow trans users the changeability, network separation and identity realness, along with the queer aspects of multiplicity, fluidity and ambiguity, needed for gender transition' (Haimson et al., 2021, p. 355). Their proposals range from redesigns for existing mainstream apps and platforms to ensure they effectively recognise fluid gender identities, to speculative designs for trans-centred technologies that promote safety, peer-support and community-building via platforms, games and social spaces.

Similarly, while the fields of sextech and femtech are overshadowed by large companies and global data markets (Wilken, Burgess and Albury, 2019), there are many smaller players in the scene seeking to build apps and platforms based on principles of design justice, a community-led design approach 'that aims explicitly to challenge, rather than reproduce, structural inequalities' (Costanza-Chock, 2020).[27] These new models (such as the escort platform Tryst, discussed in Chapter 4) privilege transparent communication regarding the systems that drive visibility and engagement. They also offer clear descriptions of when, how and why data is collected and utilised, and they deploy community moderation guidelines that do not police 'suggestiveness' in images and text, but focus instead on user safety and security. For example, web-based social app Lips Social invites users to post sexually explicit content, but uses a 'consent-based' tagging system, co-designed with queer young people, to moderate user feeds. Users are able to choose between an auto-generated feed, or a manual version, which allows them to see certain kinds of content, while also nominating specific tags they do not want to see (Iovine, 2021).

There is also hope, we argue, in the quiet, practical work of multiple non-profit and non-governmental organisations that seek to de-commodify personal data and look for its potential to achieve public good *with* the people who might otherwise be left behind. As some of the spaces where this potential plays out, we have pointed to the moments of intimate interaction with data, the situated instances of data management and data making or making do involved in data literacies, and the collectives that form and mobilise on the basis of their data interactions. While we make no heroic claims about these initiatives, we do think they provide 'glimmers and offers of hope' (Littler, 2016) for more humanistic and responsible data futures.

Groups working to establish data collaboration and co-ops offer promising cases of how everyday social data may be mobilised for public good. This might include groups like the Salus Co-op in Spain who are forging new processes for trusted data sharing that enable citizens to control their health data while facilitating sharing to improve collective health outcomes.[28] Other initiatives are developing the architecture for sovereign personal data licences, identity management and control (see for example Sovrin, DECODE and MIDATA in Europe, and the long-running National Neighborhood Indicators Partnership [NNIP] led by the Urban Institute in the United States).[29] While many community-oriented data initiatives have demonstrated the potential for enabling a wider range of voices and experiences that have often been marginalised through the collection and use of data, they also show their limits. These can be bespoke and transient movements or groups coming together while there is a pressing need, coupled with the availability of expertise and funding (often philanthropic). What we have been pointing to throughout this book is a wider kind of hope in data futures embedded in everyday lives, contexts and practices. We are advocating for attention to and facilitation of agency-building amidst the ebb and flow of everyday data cultures. This is to emphasise through research and action the material and situated responses that may be possible through the things people do with data.

The everyday data cultures project is wide open. Its valences are not dictated solely by Big Tech and the expanding reach of their platforms, cloud services, integrated databases and auto machine learning packages – after all, data is not a finite resource that these players get to solely own and control. But at the same time, we are far from finding the right language and the most effective mechanisms for plotting a way through the many concerns raised by critical data studies, while at the same time retaining a space for everyday practices of play, expression and pleasure.

Equally, locating hope solely in the role of (better) government regulation, industry self-regulation, or technological fixes even where they are guided by data privacy advocates, risks alienating by alternative means the people affected by the ordinary contexts of data production and use. It is vitally important that such efforts do not rely on a deficit model of an ignorant, easily manipulated populace; or, on the other hand, of an innocent mass of ordinary people – ordinary people can engage in harmful data practices, too.

This means that data ethics principles might need to take into account more thoughtful and careful data practices on the part of consumers, users and citizens, as well as on the part of platforms, governments and organisations, and for technologists to draw on these principles and consider the roles their technologies might play 'downstream', in exacerbating or ameliorating toxic cultures of use.

There are emerging models of data citizenship that could fit our brief for finding spaces and resources of hope. These need to be more than about data governance, even though governance, regulation, agreements and contracts are vital. Hope is sparked where they are built on respect for everyday practice and vernacular knowledges, including the knowledges of those who have been historically marginalised, or are newly marginalised by data-driven systems; their initiatives would respect agency, experience, locality and cultural sensitivities or lore. They would be situated, particular, dynamic and responsive. Effective responses to datafication both build on and supplement individual experiences and injustices in order to address the issues posed by datafication at a larger social or collective level.

Here we want to highlight work on Indigenous data sovereignty as a key space of hope (Kukutai and Taylor, 2016). It is often noted that Indigenous people have been and continue to be over-surveilled (watched, studied and policed) but under-counted (and thereby under-represented and unheard). Data is as important to the sovereignty of a people as language, artefacts, landmarks, beliefs and cultural knowledge and natural resources. Self-determination with and through data begins with 'disaggregation' – or breaking down the hierarchies of population counts to account for the particular circumstances, histories and lived experiences of colonised and marginalised people. As Tahu Kukutai and John Taylor eloquently argue, 'missing from those conversations have been the inherent and inalienable rights and interests of indigenous peoples relating to the collection, ownership and application of data about their people, lifeways and territories' (2016, p. 2). Indigenous ways of knowing can offer new models for data stewardship that are built on collaborative rather than individual or proprietary responsibility, and more respectful forms of consent. Addressing Indigenous data sovereignty can offer a better way forward for all people.

Alternative models of data governance are being enacted somewhat experimentally around the globe, usually operating through local municipal contexts within varied regulatory jurisdictions. The goals of open government data have for some time been about making the data collected through public service available for use as a public good. While they seem to be opposing forces, data sharing and sovereignty (or agency, control and consent) are the basis of recent attempts to steer away from the 'colonising' or private property approaches to data that dominate the business plans of start-ups and Big Tech.

Experimental models of data management and use are being built around local community partnerships and aim to incorporate rights-based principles (Micheli et al., 2020). The short history of open data – following other forms of open such as open-source software, and FAIR (findable, accessible, interoperable, reusable) data principles – has attracted many critics along the way (Jethani and Leorke, 2021). In an early agenda setting paper, Michael Gurstein (2003), for instance, detailed the difference between principles and action, pointing to the limitations of open *access*, at the expense of enabling effective, community-led effective data *use*. Approaches grounded in everyday data cultures might take into account the vast heterogeneity of data use that has since unfolded, and look to amplify its individual, social and cultural benefits.

Implications for research practice

Some years ago, the UK-based Open Air Laboratories (Opal), which is coordinated by Imperial College, formed a partnership with Forest Research (an arm of the UK Forestry Commission) and the Food and Environment Research Agency to launch the Opal Tree Health Survey in mid-2013 (Kinver, 2020). This was a project that sought to obtain 'crowd-sourced' tree health data. The Opal Survey is regarded as a pioneering 'citizen science' project. Over the ensuing six years, 39,000 people took part – 80 per cent of whom are reported to have had no prior experience working with trees – surveying trees in streets, schools, parks and gardens across the UK (Kinver, 2020). While questions remain about the 'robustness' of the data collected and its usefulness to research scientists (repeating the tensions between 'known and knowing publics'), the project has nevertheless been regarded as a success, not only in terms of the geographical distribution of the survey results, but also and especially in raising public awareness about tree health and for the level of citizen engagement in public data collection (Kinver, 2020).

What would such approaches look like for digital humanities and social science research, as our everyday lives and media environments become increasingly entangled with data and automation? We are already involved in a suite of data donation-based 'citizen science' projects, where we turn the tables on some of the major digital platforms by involving the public in collecting the digital traces of platforms' data operations in their own lives – including, for example, Facebook's targeted advertising practices (Burgess et al., 2021). We think that user-contributed data, when combined with thoughtful qualitative research, can contribute to both public oversight and awareness of platforms' data operations – one of the goals, surely, of critical data studies – while at the same time enabling meaningful conversations with participants that help to build data literacies and shared principles for everyday data ethics.

In our own work, we have explored the potential of participatory methods for investigating everyday data cultures in collaboration with the people whose practices shape and are shaped by them. For example, we have drawn on social science methodologies such as data journeys (Bates, Lin and Goodale, 2016) and data walks (Murray, Falkenburger and Saxena, 2015) in collaboration with workers in non-profit organisations, in order to build data capacities and capabilities (Yao et al., 2021), as well as show-and-tell approaches for understanding everyday individualised data management practices (Wilken and Kennedy, 2022). We've engaged with creative methods to explore dating app users' experiences of safety (Albury et al., 2021); and adapted transfeminist and design justice-based methodologies with technologists and sextech entrepreneurs, in order to rethink the connections between data, sexuality and gender (Costanza-Chock, 2020; Varon, 2020).

As we have engaged with multiple data stakeholders (including policy-makers, health workers, dating app users, sextech entrepreneurs and non-profit staffers) we have been continually surprised and challenged to extend our theoretical or critical understandings of data cultures to account for the messiness and complexity of the everyday. This has prompted multiple shifts in our own understandings of data cultures – for example in Wilken's empirical research with allied health practitioners that looked at the persistence of the use of fax machines in the era of digital, encrypted communication and cloud data management. While senior bureaucrats attributed this loyalty to a legacy technology that is believed to offer data security, front-line health workers used the language of familiarity, habit and routine. By bringing the tensions between these grounded accounts of technological change into conversation with critical digital and data studies, we gain new opportunities for understanding the specific and contextual ways that data cultures emerge and evolve.

Finally, as we know from our own work, research grounded in everyday data cultures can be enjoyable, and, in the right situation, downright fun (Carlson and Frazer, 2021). This kind of research practice can open us up to surprising and creative connections between and among people and technologies, and to the future possibilities that these relationships might help us to imagine and to shape.

Notes

1 We thank Yong Lim of Seoul National University for drawing our attention to the kinds of practices captured in this fictional vignette.
2 See, for example, this thread from the 'Not The Onion' subreddit: https://www.reddit.com/r/nottheonion/comments/oc0ckw/bay_area_police_sergeant_played_taylor_swift_to/Cabanatuan.
3 We note that a similar, but much more expansive, approach to the use of ficto-critical accounts was taken by Rob Kitchin in his book *Data Lives* (2021), which traverses some of the same themes as we do here.
4 The Owner blog's guide to Instagram for restaurants is at https://owner.com/the-ultimate-guide-to-instagram-for-restaurant-professionals.
5 For T. S. Eliot, too, 'the culture of the individual is dependent upon the culture of a group or class, and [...] the culture of the group or class is dependent upon the culture of the whole society to which that group or class belongs'. It is the last in this list – the 'whole way of life' or, as he refers to it, the 'culture of the whole society' (1948, p. 21) – that is 'fundamental' and should form

our principal object of focus. However, for early British cultural studies scholars, all three of these multiple scales or levels of culture were regarded as important, collectively forming a complex system of meaning-making practices, common to all people rather than the preserve of an elite, but at the local level varied and patterned according to place and class.

6 OnlyFans is a social-media-style app that facilitates payment in exchange for messaging and images. It is mainly (but not exclusively) used as a platform for sharing NSFW (Not Safe For Work) content, ranging from sexy selfies to explicit pornographic videos. Camming (or webcamming) platforms such as Chaturbate enable payment for live-streamed sexual content.

7 The FemTech Collective website can be found at https://femtechcollective.com.

8 The website for the Lioness is at https://lioness.io.

9 The Lioness blog is at https://lioness.io/blogs/lioness-product-updates/live-view-a-new-way-to-explore-with-lioness.

10 Lelo's F1 Developer kit is described at https://www.lelo.com/f1s-v2.

11 Miss Ruby's F1 review is found at https://missrubyreviews.com/review-lelo-f1s-developers-kit-red-penis-masturbator/lelo-f1s-review-by-miss-ruby-reviews-1.

12 Lioness reviews are found at https://lioness.io/pages/reviews#shopify-product-reviews.

13 The term BDSM is generally understood as an acronym for bondage and discipline, plus dominance and submission and/or sado-masochism.

14 The BDSM checklist is at http://www.wikiphilia.net/bdsmchecklist/bdsmchecklist.php.

15 The Human Sex Map is at https://humansexmap.com.

16 The Secureteen website is at https://www.secureteen.com.

17 @NotLaja, 'Lil Nas is a Genius, A thread', at https://twitter.com/NotLaja/status/1143893458245095425.

18 'Stan' is a portmanteau of 'stalker' and 'fan', and denotes a particularly avid, very online, mode of fandom.

19 Salty, an investigation into algorithmic bias in content policing on Instagram, can be found at https://saltyworld. net/algorithmicbiasreport-2.
20 The work of the activist group who produced a series of images and versions of the EveryBODYVisible logo can be found at https://everybodyvisible.com.
21 On Facebook's moderation policies concerning 'adult content', see https://www.facebook.com/policies/ads/ prohibited_content/adult_content.
22 On the Adult Performers Actors Guild (APAG) and their activist work on Instagram discrimination, see https:// apagunion.com/2019/04/18/instagram-discrimination.
23 For Tarleton Gillespie on content suppression, see https:// threadreaderapp.com/thread/1387170486988247042. html.
24 For more information about Flowing Data, see https:// flowingdata.com.
25 The Art/Data/Health website is at https://www.artdata-health.org.
26 For more information about the Living With Data project, see https://livingwithdata.org.
27 On the building of apps and platforms based on principles of design justice, see https://design-justice.pubpub.org.
28 On the Salus Co-op, see https://www.saluscoop.org/ ?lang=en.
29 On the work of Sovrin, see https://sovrin.org. For information on DECODE, see https://decodeproject.eu. On the MIDATA Co-op, see https://www.midata.coop/en/home. On the National Neighborhood Indicators Partnership, or NNIP, in the United States, see https://www.neighbor hoodindicators.org.

References

Abend, Pablo and Mathias Fuchs (2016). Introduction: the quantified self and statistical bodies. *Digital Culture and Society*, 2(1), pp. 5–21.

Abidin, Crystal (2021). From 'networked publics' to 'refracted publics': a companion framework for researching 'below the radar' studies. *Social Media + Society*, 7(1), https://doi.org/10.1177/2056305120984458.

Ahmed, Alex A. (2018). Trans competent interaction design: a qualitative study on voice, identity, and technology. *Interacting with Computers*, 30(1), pp. 53–71.

Albury, Kath (2017). Just because it's public doesn't mean it's any of your business: adults' and children's sexual rights in digitally mediated spaces. *New Media & Society*, 19(5), pp. 713–25.

Albury, Kath, Amir Aryani, Jane Farmer, James Kelly, Anthony McCosker, Sandun Silva, Julie Tucker and Jihoon Woo (2021). *Data for Good Collaboration*. Melbourne: Swinburne University of Technology, https://apo.org.au/node/312198.

Albury, Kath, Anthony McCosker and Clifton Evers (2021). Men seeking women: awkwardness, shame, and other affective encounters with dating apps. *First Monday*, https://doi.org/10.5210/fm.v26i4.11637.

Albury, Kath, Anthony McCosker, Tinonee Pym and Paul Byron (2020). Dating apps as public health 'problems': cautionary tales and vernacular pedagogies in news media. *Health Sociology Review*, 29(3), pp. 232–48.

Albury, Kath, Christopher Dietzel, Tinonee Pym, Son Vivienne and Teddy Cook (2021). Not your unicorn: trans dating app users' negotiations of personal safety and sexual health. *Health Sociology Review*, 30(1), pp. 72–86.

Albury, Kath, Jean Burgess, Ben Light, Kane Race and Rowan Wilken (2017). Data cultures of mobile dating and hook-up apps: emerging issues for critical social science research. *Big Data & Society*, https://doi.org/10.1177/2053951717720950.

Albury, Kath and Kate Crawford (2012). Sexting, consent and young people's ethics: beyond *Megan's Story*. *Continuum*, 26(3), pp. 463–73.

Albury, Kath and Paul Byron (2016). Safe on my phone? Same-sex attracted young people's negotiations of intimacy, visibility, and risk on digital hook-up apps. *Social Media + Society*, 2(4), https://doi.org/10.1177/2056305116672887.

Albury, Kath, Paul Byron, Anthony McCosker, Tinonee Pym et al. (2019). *Safety, Risk and Wellbeing on Dating Apps – Research Report*. Melbourne: Swinburne University of Technology.

Altheide, David and Robert P. Snow (1979). *Media Logic*. Beverly Hills: Sage.

Amoore, Louise (2020). *Cloud Ethics: Algorithms and the Attributes of Ourselves and Others*. Durham, NC: Duke University Press.

Andersen, Jack (2018). Archiving, ordering, and searching: search engines, algorithms, databases, and deep mediatization. *Media, Culture & Society*, 40(9), pp. 1135–50.

Andrejevic, Mark (2010). Social network exploitation. In Z. Papacharissi (ed.), *A Networked Self: Identity, Community, and Culture on Social Network Sites*. New York: Routledge, pp. 82–101.

Andrejevic, Mark (2019a). *Automated Media.* New York: Routledge.

Andrejevic, Mark (2019b). Automating surveillance. *Surveillance & Society*, 17(1/2), pp. 7–13.

Andrejevic, Mark, Hugh Davies, Ruth DeSouza, Larissa Hjorth and Ingrid Richardson (2021). Situating 'careful surveillance'. *International Journal of Cultural Studies*, 24(4), pp. 567–83.

Anwar, Mohammad A. and Mark Graham (2020). Hidden transcripts of the gig economy: labour agency and the new art of resistance among African gig workers. *Environment and Planning A: Economy and Space*, 52(7), pp. 1269–91.

Are, Carolina (2019). Instagram denies censorship of pole dancers and sex workers. *Blogger on Pole*, 23 July, https://bloggeronpole.com/2019/07/instagram-denies-censorship-of-pole-dancers-and-sex-workers.

Are, Carolina (2020). How Instagram's algorithm is censoring women and vulnerable users but helping online abusers. *Feminist Media Studies*, 20(5), pp. 741–4.

Are, Carolina (2021). The Shadowban Cycle: an autoethnography of pole dancing, nudity and censorship on Instagram. *Feminist Media Studies*, https://doi.org/10.1080/14680777.2021.1928259.

Arnold, Matthew (1869). *Culture and Anarchy: An Essay in Political and Social Criticism.* London: Smith, Elder & Co.

Asher Hamilton, Isobel (2020). 77 cell towers have been set on fire so far due to a weird coronavirus 5G conspiracy theory. *Business Insider Australia*, 7 May, https://www.businessinsider.com.au/77-phone-masts-fire-coronavirus-5g-conspiracy-theory-2020-5.

Athique, Adrian (2018). The dynamics and potentials of big data for audience research. *Media, Culture & Society*, 40(1), pp. 59–74.

Attig, Christiane and Thomas Franke (2020). Abandonment of personal quantification: a review and empirical study investigating reasons for wearable activity tracking attrition. *Computers in Human Behavior*, 102, pp. 223–37.

Aubrey, Sophie (2021). Dating often leaves Alyssa feeling

degraded, and she's fed up. *Sydney Morning Herald*, 14 June, https://www.smh.com.au/lifestyle/life-and-relationships/dating-often-leaves-alyssa-feeling-degraded-and-she-s-fed-up-20210611-p5809o.html.

Bailey, Jane, Asher Flynn and Nicola Henry (eds.) (2021). *The Emerald International Handbook of Technology-Facilitated Violence and Abuse*. Bingley: Emerald.

Bandura, Albert (1969). Social-learning theory of identificatory processes. In D. A. Goslin (ed.), *Handbook of Socialization Theory and Research*. Chicago: Rand McNally & Company, pp. 213–62.

Barreneche, Carlos (2012). Governing the geocoded world: environmentality and the politics of location platforms. *Convergence: The International Journal of Research into New Media Technologies*, 18(3), pp. 331–51.

Barreneche, Carlos (2015). The cluster diagram: a topological analysis of locative networking. In R. Wilken and G. Goggin (eds.), *Locative Media*. New York: Routledge, pp. 107–17.

Barreneche, Carlos and Rowan Wilken (2015). Platform specificity and the politics of location data extraction. *European Journal of Cultural Studies*, 18(4–5), pp. 497–513.

Barthes, Roland (1993). *Mythologies*, trans. A. Lavers. London: Vintage.

Barton, David, Mary Hamilton and Roz Ivanič (eds.) (2000). *Situated Literacies: Reading and Writing in Context*. Hove: Psychology Press.

Bassett, Caroline, Aristea Fotopoulou and Kate Howland (2015). Expertise: a report and a manifesto. *Convergence*, 21(3), pp. 328–42.

Bates, Jo, Yu-Wei Lin and Paula Goodale (2016). Data journeys: capturing the socio-material constitution of data objects and flows. *Big Data & Society*, https://doi.org/10.1177/2053951716654502.

Bauman, Zygmunt and David Lyon (2012). *Liquid Surveillance: A Conversation*. Cambridge: Polity.

BBC News (2021). Face editing: Japanese biker tricks internet into thinking he is a young woman. *BBC News*, 19 March, https://www.bbc.com/news/world-asia-56447357.

Becker, Karin (1995). Media and the ritual process. *Media, Culture & Society*, 17(4), pp. 629–46.

Bell, Catherine M. (2009). *Ritual Theory, Ritual Practice*. New York: Oxford University Press.

Benbunan-Fich, Raquel (2017). The ethics of online research with unsuspecting users: from A/B testing to C/D experimentation. *Research Ethics*, 13(3–4), pp. 200–18.

Benjamin, Ruha (2019). *Race After Technology: Abolitionist Tools for the New Jim Code*. Cambridge: Polity.

Beraldo, Davide and Stefania Milan (2019). From data politics to the contentious politics of data. *Big Data & Society*, 6(2), https://doi.org/10.1177/2053951719885967.

Berlant, Lauren (1998). Intimacy: a special issue. *Critical Inquiry*, 24 (Winter), pp. 281–8.

Bishop, Sophie (2019). Managing visibility on YouTube through algorithmic gossip. *New Media & Society*, 21(11–12), pp. 2589–606.

Bivens, Rena (2017). The gender binary will not be deprogrammed: ten years of coding gender on Facebook. *New Media & Society*, 19(6), pp. 880–98.

Blanchot, Maurice (1987). Everyday speech, trans. S. Hanson. *Yale French Studies*, 73, pp. 12–20.

Blunt, Danielle and Ariel Wolf (2020). *Erased: The Impact of FOSTA-SESTA and the Removal of Backpage*. Hacking//Hustling.org, https://hackinghustling.org/wp-content/uploads/2020/02/Erased_Updated.pdf.

Blunt, Danielle, Emily Coombes, Shanelle Mullin and Ariel Wolf (2020). *Posting into the Void*. Hacking//Hustling.org, https://hackinghustling.org/wp-content/uploads/2020/09/Posting-Into-the-Void.pdf.

Bosely, Matilda (2021). TikTok accidentally detected my ADHD. For 23 years everyone missed the warning signs. *Guardian*, 4 June, https://www.theguardian.com/commentisfree/2021/jun/04/tiktok-accidentally-detected-my-adhd-for-23-years-everyone-missed-the-warning-signs.

Bossio, Diana and Anthony McCosker (2021). Reluctant

selfies: older people, social media sharing and digital inclusion. *Continuum*, 35(4), pp. 634–47.

boyd, danah (2010). Social network sites as networked publics: affordances, dynamics, and implications. In Z. Papacharissi (ed.), *A Networked Self: Identity, Community, and Culture on Social Network Sites*. New York: Routledge, pp. 39–58.

boyd, danah (2012). How parents normalized teen password sharing. *Zephoria*, 23 January, https://www.zephoria.org/thoughts/archives/2012/01/23/how-parents-normalized-teen-password-sharing.html.

boyd, danah (2018). You think you want media literacy … do you? *Medium*, 10 March, https://points.datasociety.net/you-think-you-want-media-literacy-do-you-7cad6af18ec2.

boyd, danah and Kate Crawford (2012). Critical questions for big data: provocations for a cultural, technological, and scholarly phenomenon. *Information, Communication & Society*, 15(5), pp. 662–79.

Bradshaw, Tim (2021). ByteDance starts selling AI that powers TikTok to other companies. *Financial Times*, 4 July, https://www.ft.com/content/bed7cba1-db7a-49c7-9d57-06fd19e14e10.

Brock, André (2020). *Distributed Blackness: African American Cybercultures*. New York: New York University Press.

Broniatowski, David A. et al. (2018). Weaponized health communication: Twitter bots and Russian trolls amplify the vaccine debate. *American Journal of Public Health*, 108(10), pp. 1378–84.

Bruns, Axel and Jean Burgess (2015). Twitter hashtags from *ad hoc* to calculated publics. In N. Rambukkana (ed.), *Hashtag Publics: The Power and Politics of Discursive Networks*. New York: Peter Lang, pp. 13–27.

Bruns, Axel, Stephen Harrington and Edward Hurcombe (2020). 'Corona? 5G? Or both?': the dynamics of COVID-19/5G conspiracy theories on Facebook. *Media International Australia*, 177(1), pp. 12–29.

Bucher, Taina (2017). The algorithmic imaginary: exploring

the ordinary affects of Facebook algorithms. *Information, Communication & Society*, 20(1), pp. 30–44.

Bucher, Taina (2018). *If ... Then: Algorithmic Power and Politics*. New York: Oxford University Press.

Bucher, Taina (2021). *Facebook*. Cambridge: Polity.

Buckingham, David (2013). *Media Education: Literacy, Learning and Contemporary Culture*. Hoboken: John Wiley & Sons.

Budin-Ljøsne, Isabelle, Harriet J. Teare, Jane Kaye and Stephen Beck et al. (2017). Dynamic consent: a potential solution to some of the challenges of modern biomedical research. *BMC Medical Ethics*, 18(1), https://doi.org/10.1186/s12910-016-0162-9.

Burgess, Jean (2006). Hearing ordinary voices: cultural studies, vernacular creativity, and digital storytelling. *Continuum: Journal of Media & Cultural Studies*, 20(2), pp. 201–14.

Burgess, Jean (2015). From 'broadcast yourself' to 'follow your interests': making over social media. *International Journal of Cultural Studies*, 18(3), pp. 281–5.

Burgess, Jean (2017a). Hookup apps' vernacular data cultures. Presentation at *The Social Life of Data*, 28 April, Melbourne, Australia, https://eprints.qut.edu.au/209731.

Burgess, Jean (2017b). Convergence. In L. Ouelette and J. Gray (eds.), *Keywords in Media Studies*. New York: New York University Press, p. 47.

Burgess, Jean, Daniel Angus, Nicholas Carah, Mark Andrejevic et al. (2021). Critical simulation as hybrid digital method for exploring the data operations and vernacular cultures of visual social media platforms, https://doi.org/10.31235/osf.io/2cwsu.

Burgess, Jean and Joshua Green (2009). *YouTube: Online Video and Participatory Culture*. Hoboken: John Wiley & Sons.

Burgess, Jean and Nancy Baym (2020). *Twitter: A Biography*. New York: New York University Press.

Burgess, Jean, Peta Mitchell and Felix Münch (2019). Social media rituals: the uses of celebrity death in digital culture.

In Z. Papacharissi (ed.), *A Networked Self and Birth, Life, Death*. New York: Routledge, pp. 224–39.

Burke, Caroline (2020). Ever spent hours on TikTok without realizing it? The app is trying to fix that. *Bustle*, 20 February, https://www.bustle.com/p/tiktoks-new-screen-time-prompts-remind-users-to-take-a-breather-21816527.

Burrell, Jenna (2016). How the machine 'thinks': understanding opacity in machine learning algorithms. *Big Data & Society*, 3(1), https://doi.org/10.1177/2053951715622512.

Cabanatuan, Michael (2021). Why did an Alameda County deputy play a Taylor Swift song while arguing with protesters? *San Francisco Chronicle*, 1 July, https://www.sfchronicle.com/bayarea/article/Why-did-an-Alameda-deputy-play-a-Taylor-Swift-16288542.php.

Carey, James W. (2009 [1989]). *Communication as Culture: Essays on Media and Society*. Revised edition. New York: Routledge.

Carey, James W. and John J. Quirk (1970). The mythos of the electronic revolution. *The American Scholar*, 39(3), pp. 395–424.

Carlson, Bronwyn (2020). Love and hate at the cultural interface: Indigenous Australians and dating apps. *Journal of Sociology*, 56(2), pp. 133–50.

Carlson, Bronwyn and Ryan Frazer (2021). *Indigenous Digital Life: The Practice and Politics of Being Indigenous on Social Media*. London: Palgrave Macmillan.

Cave, Damien (2021). 'The most unsafe workplace'? Parliament, Australian women say. *New York Times*, 5 April, https://www.nytimes.com/2021/04/05/world/australia/parliament-women-rape-metoo.html.

Chambers, Deborah (2017). Networked intimacy: algorithmic friendship and scalable sociality. *European Journal of Communication*, 32(1), pp. 26–36.

Chang, Hasok (2004). *Inventing Temperature: Measurement and Scientific Progress*. Oxford: Oxford University Press.

Cheney-Lippold, John (2017). *We Are Data: Algorithms and*

the Making of Our Digital Selves. New York: New York University Press.

Chow, Andrew R. (2019). The Yeehaw Agenda is about more than cowboy hats. It's about American identity. *Time*, 21 November, https://time.com/5735430/yeehaw-agenda-black-artists-reclaiming-cowboy-image.

Clegg, Nick (2021). You and the algorithm: it takes two to tango. *Medium*, 1 April, https://nickclegg.medium.com/you-and-the-algorithm-it-takes-two-to-tango-7722b19aa1c2.

Clement, Tanya and Amelia Acker (2019). Data cultures, culture as data. Special issue of *Cultural Analytics*, https://doi.org/10.22148/16.035.

Clucas, Stephen (2000). Cultural phenomenology and the everyday. *Critical Quarterly*, 41(1), pp. 8–34.

Comella, Lynne (2017). *Vibrator Nation: How Feminist Sex Toy Stores Changed the Business of Pleasure*. Durham, NC: Duke University Press.

Constine, Josh (2019). Instagram now demotes vaguely 'inappropriate' content. *Tech Crunch*, 11 April, https://techcrunch.com/2019/04/10/instagram-borderline.

Costanza-Chock, Sasha (2020). *Design Justice: Community-Led Practices to Build the Worlds We Need*. Cambridge, MA: MIT Press.

Cotter, Kelly (2019). Playing the visibility game: how digital influencers and algorithms negotiate influence on Instagram. *New Media & Society*, 21(4), pp. 895–913.

Couldry, Nick and Andreas Hepp (2017). *The Mediated Construction of Reality*. Cambridge: Polity.

Couldry, Nick and Ulises A. Mejias (2019). *The Costs of Connection: How Data is Colonizing Human Life and Appropriating it for Capitalism*. Stanford: Stanford University Press.

Crawford, Kate (2021). *Atlas of AI*. New Haven: Yale University Press.

Crawford, Kate and Trevor Paglen (2019). Excavating AI: the politics of images in machine learning training sets. *Excavating AI*, https://excavating.ai.

Cresswell, Tim (2006). *On the Move: Mobility in the Modern Western World*. New York: Routledge.

Dalton, Paul (2021). The Acta of William the Conqueror, Domesday Book, the Oath of Salisbury, and the legitimacy and stability of the Norman Regime in England. *Journal of British Studies*, 60, pp. 29–65.

Damkaer, Maya, Claire Southerton and Anders Albrechtslund (2019). Relief from communication: parental surveillance technologies, trust and care. Selected Papers of AoIR #2019: The 20th Annual Conference of the Association of Internet Researchers. Brisbane, Australia, 2–5 October.

David, Gaby and Carolina Cambre (2016). Screened intimacies: Tinder and the swipe logic. *Social Media + Society*, 2(2), https://doi.org/10.1177/2056305116641976.

Dayan, Daniel and Elihu Katz (1994). *Media Events: The Live Broadcasting of History*. Cambridge, MA: Harvard University Press.

De Certeau, Michel (1984). *The Practice of Everyday Life*, trans. S. Rendall. Berkeley: University of California Press.

De Certeau, Michel, Luce Giard and Pierre Mayol (1998). *The Practice of Everyday Life, Volume 2: Living and Cooking*, trans. T. J. Tomasik. Minneapolis: University of Minnesota Press.

De Ridder, Sander (2021). The datafication of intimacy: mobile dating apps, dependency and everyday life. *Television & New Media*, https://doi.org/10.1177/15274764211052660.

Denton, Emily, Alex Hanna, Razvan Amironesei, Andrew Smart and Hilary Nicole (2021). On the genealogy of machine learning datasets: a critical history of ImageNet. *Big Data & Society*, https://doi.org/10.1177/20539517211035955.

Dewey, John (2016 [1927]). *The Public and its Problems: An Essay in Political Inquiry*. Athens: Swallow Press.

D'Ignazio, Catherine and Lauren F. Klein (2020). *Data Feminism*. Cambridge, MA: MIT Press.

D'Ignazio, Catherine and Rahul Bhargava (2015). Approaches to building big data literacy. *Bloomberg Data for Good Exchange Conference*, https://www.media.mit. edu/publications/approaches-to-building-big-data-literacy.

Dhaenens, Frederik and Jean Burgess (2019). 'Press play for pride': the cultural logics of LGBTQ-themed playlists on Spotify. *New Media & Society*, 21(6), pp. 1192–211.

Dobson, Amy Shields, Brady Robards and Nicholas Carah (eds.) (2018). *Digital Intimate Publics and Social Media*. London: Palgrave Macmillan.

Dobson, Amy Shields and Jessica Ringrose (2016). Sext education: pedagogies of sex, gender and shame in the schoolyards of Tagged and Exposed. *Sex Education*, 16(1), pp. 8–21.

Dragiewicz, Molly, Jean Burgess, Ariadna Matamoros-Fernández, Michael Salter, Nicolas P. Suzor, Delanie Woodlock and Bridget Harris (2018). Technology facilitated coercive control: domestic violence and the competing roles of digital media platforms. *Feminist Media Studies*, 18(4), pp. 609–25.

Dragiewicz, Molly, Patrick O'Leary, Jeffrey Ackerman, Christine Bond, Ernest Foo, Amy Young and Claire Reid (2020). Children and technology-facilitated abuse in domestic and family violence situations. Full Report, December 2020. Office of the eSafety Commissioner, Sydney, https://www.esafety.gov.au/research/children-and-technology-facilitated-abuse-domestic-and-family-violence-situations.

D-River (2019). App to help juggle Uber Didi and Ola?. *UberPeople Forums*, 6 July, https://www.uberpeople.net/threads/app-to-help-juggle-uber-didi-and-ola.338327/#post-5165966.

Duffy, Erin Brooke (2018). *(Not) Getting Paid to Do What You Love*. New Haven: Yale University Press.

Duguay, Stefanie (2020). You can't use this app for that: exploring off-label use through an investigation of Tinder. *The Information Society*, 36(1), pp. 30–42.

Edwards, Paul N. (2010). *A Vast Machine: Computer Models, Climate Data, and the Politics of Global Warming*. Cambridge, MA: MIT Press.

Eliot, T. S. (1948). *Notes Towards the Definition of Culture*. London: Faber & Faber.

Ely, Margot, Ruth Vinz, Margret Downing and Maryanne Anzul (1997). *On Writing Qualitative Research: Living by Words*. London: Falmer Press.

Eriksson, Maria and Anna Johansson (2017). 'Keep smiling!' Time, functionality and intimacy in Spotify's featured playlists. *Cultural Analysis*, 16(1), pp. 67–82.

Eriksson, Maria, Rasmus Fleischer, Anna Johansson, Pelle Snickars and Patrick Vonderau (2019). *Spotify Teardown: Inside the Black Box of Streaming Music*. Cambridge, MA: MIT Press.

Esmonde, Katelyn (2019). Training, tracking, and traversing: digital materiality and the production of bodies and/in space in runners' fitness tracking practices. *Leisure Studies*, 38(6), pp. 804–17.

Eubanks, Virginia (2018). *Automating Inequality: How High-Tech Tools Profile, Police, and Punish the Poor*. New York: St. Martin's Press.

Evans, Will (2019). Ruthless quotas at Amazon are maiming employees. *The Atlantic*, 26 November, https://www.theatlantic.com/technology/archive/2019/11/amazon-warehouse-reports-show-worker-injuries/602530.

Felski, Rita (2000). *Doing Time: Feminist Theory and Postmodern Culture*. New York: New York University Press.

Fiore-Gartland, Brittany and Gina Neff (2015). Communication, mediation, and the expectations of data: data valences across health and wellness communities. *International Journal of Communication*, 9, https://ijoc.org/index.php/ijoc/article/view/2830.

Fisher, Max (2021). 'Belonging is stronger than facts': the age of misinformation. *New York Times*, 7 May, https://www.nytimes.com/2021/05/07/world/asia/misinformation-disinformation-fake-news.html.

Fiske, John and John Hartley (1978). *Reading Television*. London: Methuen.

Flore, Jacinth and Kieran Pienaar (2020). Data-driven intimacy: emerging technologies in the (re)making of sexual subjects and 'healthy' sexuality. *Health Sociology Review*, 29(3), pp. 279–93.

Flyverbom, Mikkel and John Murray (2018). Datastructuring – organizing and curating digital traces into action. *Big Data & Society*, 5(2), https://doi.org/10.1177/2053951718799114.

Ford, Brody (2021). DoorDash drivers game algorithm to increase pay. *Bloomberg*, 6 April, https://www.bloomberg.com/news/articles/2021–04–06/doordash-workers-are-trying-to-game-the-algorithm-to-increase-pay.

Fotopoulou, Aristea (2021). Conceptualising critical data literacies for civil society organisations: agency, care, and social responsibility. *Information, Communication & Society*, 24(11), pp. 1640–57.

Foucault, Michel (1990 [1987]). *The History of Sexuality Volume One: An Introduction*. New York: Vintage Books.

Fourcade, Marion and Fleur Johns (2020). Loops, ladders and links: the recursivity of social and machine learning. *Theory and Society*, 49(5), pp. 803–32.

Fox, Susannah and Maeve Duggan (2013). *Tracking for Health*. Pew Research Center's Internet and American Life Project.

Freire, Paulo (1968). *Pedagogy of the Oppressed*. Revised edition. New York: Continuum.

Fritsch, Katrin (2018). Towards an emancipatory understanding of widespread datafication. *Medium*, 12 February, https://medium.com/@katrinfritsch/towards-an-emancipatory-understanding-of-widespread-datafication-5a53ed79eb0f.

Galant, Simona (2020). Data spotlight: how Spotify Wrapped makes music data feel personable. *Springboard Blog*, 26 November, https://www.springboard.com/blog/data-science/spotify-data-insights.

Gibbs, Martin, James Meese, Michael Arnold, Bjorn Nansen and Marcus Carter (2015). #funeral and Instagram: death, social media, and platform vernacular. *Information, Communication & Society*, 18(3), pp. 255–68.

Gilbert, Ben (2019). There's a simple reason your new smart TV was so affordable: it's collecting and selling data, and serving you ads. *Business Insider Australia*, 14 January, https://www.businessinsider.com.au/smart-tv-data-collection-advertising-2019–1?r=US&IR=T.

Gillespie, Tarleton (2018). *Custodians of the Internet*. New Haven: Yale University Press.

Gillett, Rosalie, Kath Albury and Zahra Stardust (2021). NSW Police want access to Tinder's sexual assault data. Cybersafety experts explain why it's a date with disaster. *The Conversation*, 28 April, https://theconversation.com/nsw-police-want-access-to-tinders-sexual-assault-data-cybersafety-experts-explain-why-its-a-date-with-disaster-159811.

Gilster, Paul (1997). *Digital Literacy*. London: John Wiley & Sons.

Ging, Debbie and Sarah Garvey (2018). 'Written in these scars are the stories I can't explain': a content analysis of pro-ana and thinspiration image sharing on Instagram. *New Media & Society*, 20(3), pp. 1181–200.

Gitelman, Lisa (ed.) (2013). *'Raw Data' is an Oxymoron*. Cambridge, MA: MIT Press.

Goggin, Gerard, Katie Ellis and Wayne Hawkins (2019). Disability at the centre of digital inclusion: assessing a new moment in technology and rights. *Communication Research and Practice*, 5(3), pp. 290–303.

Goode, Luke (2018). Life, but not as we know it: A.I. and popular imagination. *Culture Unbound*, 10(2), pp. 185–207.

Graham, Timothy, Axel Bruns, Guangnan Zhu and Rod Campbell (2020). Like a virus: the coordinated spread of coronavirus disinformation. The Australia Institute, Canberra, A.C.T.

Gran, Anne-Britt, Peter Booth and Taina Bucher (2021). To be or not to be algorithm aware: a question of a new digital divide? *Information, Communication & Society*, 24(12), pp. 1779–96.

Gray, Jonathan, Carolin Gerlitz and Liliana Bounegru (2018). Data infrastructure literacy. *Big Data & Society*, 5(2), https://doi.org/10.1177/2053951718786316.

Gray, Mary and Siddharth Suri (2019). *Ghost Work: How*

to Stop Silicon Valley from Building a New Global Underclass. Boston: Houghton Mifflin Harcourt.

Gregg, Melissa (2018). *Counterproductive*. Durham, NC: Duke University Press.

Grossberg, Lawrence (2015). A special editorial. *Cultural Studies*, 29(2), pp. 105–7.

Gurstein, Michael (2003). Effective use: a community informatics strategy beyond the digital divide. *First Monday*, https://doi.org/10.5210/fm.v8i12.1107.

Hagen, Anja Nylund (2015). The playlist experience: personal playlists in music streaming services. *Popular Music and Society*, 38(5), pp. 625–45.

Haimson, Oliver L., Avery Dame-Griff, Elias Capello and Zahari Richter (2021). Tumblr was a trans technology: the meaning, importance, history, and future of trans technologies. *Feminist Media Studies*, 21(3), pp. 345–61.

Haimson, Oliver L., Dykee Gorrell, Denny L. Starks and Zu Weinger (2020). Designing trans technology: defining challenges and envisioning community-centered solutions. In *Proceedings of the 2020 CHI Conference on Human Factors in Computing Systems*, pp. 1–13.

Hallinan, Blake and Ted Striphas (2016). Recommended for you: the Netflix prize and the production of algorithmic culture. *New Media & Society*, 18(1), pp. 117–37.

Halpern, Orit (2015). *Beautiful Data: A History of Vision and Reason since 1945*. Durham, NC: Duke University Press.

Hao, Karen (2020). We read the paper that forced Timnit Gebru out of Google. Here's what it says. *MIT Technology Review*, 4 December, https://www.technologyreview.com/2020/12/04/1013294/google-ai-ethics-research-paper-forced-out-timnit-gebru.

Hao, Karen (2021). Inside the fight to reclaim AI from Big Tech. *MIT Technology Review*, 14 June, https://www.technologyreview.com/2021/06/14/1026148/ai-big-tech-timnit-gebru-paper-ethics.

Hartley, John (1987). Invisible fictions: television audiences, paedocracy, pleasure. *Textual Practice*, 1(2), pp. 121–38.

He, Huifeng (2020). China's truck drivers see dead end ahead amid rising costs and new toll system. *South China Morning Post*, 12 January, https://www.scmp.com/economy/china-economy/article/3045532/chinas-truck-drivers-see-dead-end-ahead-amid-rising-costs-and.

Hearn, Alex (2020). Twitter apologises for 'racist' image-cropping algorithm. *Guardian*, 21 September, https://www.theguardian.com/technology/2020/sep/21/twitter-apologises-for-racist-image-cropping-algorithm.

Helmond, Anne (2015). The platformization of the web: making web data platform ready. *Social Media + Society*, 1(2), https://doi.org/10.1177/2056305115603080.

Hepp, Andreas (2019). *Deep Mediatization*. London: Routledge.

Herbert, Daniel, Amanda D. Lotz and Aswin Punathambekar (2020). *Media Industry Studies*. Cambridge: Polity.

Highfield, Tim (2017). *Social Media and Everyday Politics*. Cambridge: Polity.

Highmore, Ben (2002). *Everyday Life and Cultural Theory: An introduction*. London: Routledge.

Highmore, Ben (2010). *Ordinary Lives: Studies in the Everyday*. London: Routledge.

Highmore, Ben (2015). *Culture*. London: Routledge.

Hintz, Arne, Lina Dencik and Karin Wahl-Jorgensen (2019). *Digital Citizenship in a Datafied Society*. Cambridge: Polity.

Hjorth, Larissa, Sarah Pink and Heather A. Horst (2018). Being at home with privacy: privacy and mundane intimacy through same-sex locative media practices. *International Journal of Communication*, 12, pp. 1209–27.

Hoggart, Richard (1957). *The Uses of Literacy*. Fair Lawn: Essential Books.

Hubel, Olaf (2017). What's new in Excel – Ignite 2017 announcements. *Microsoft*, 16 October, https://techcommunity.microsoft.com/t5/excel/what-is-new-in-excel-ignite-2017-announcements/m-p/117029.

Humphreys, Lee (2018). *The Qualified Self: Social Media and the Accounting of Everyday Life*. Cambridge, MA: MIT Press.

Hunt, Lola (2021). Beginner's guide to Tryst analytics and algorithms! *Tryst Blog*, 22 February, https://tryst.link/blog/beginners-guide-to-tryst-analytics.

Hwang, Tim (2020). *Subprime Attention Crisis: Advertising and the Time Bomb at the Heart of the Internet*. New York: FSG Originals x Logic.

Iliadis, Andrew and Federica Russo (2016). Critical data studies: an introduction. *Big Data & Society*, 3(2), https://doi.org/10.1177/2053951716674238.

Iovine, Anna (2021). Lips is a new social network where sexual expression is welcome. *Mashable South East Asia*, 26 January, https://sea.mashable.com/culture/14226/lips-is-a-new-social-network-where-sexual-expression-is-welcome.

Iqbal, Nosheen (2021). Celebrity buzz: how stars' bedroom toys have got us all talking about sex. *Guardian*, 13 June, https://www.theguardian.com/lifeandstyle/2021/jun/13/celebrity-buzz-how-stars-bedroom-toys-have-got-us-all-talking-about-sex.

Jagose, Annamarie (2013). *Orgasmology*. Durham, NC: Duke University Press.

Jansen, Sue Curry and Jefferson Pooley (2021). Blurring genres and violating guild norms: a review of reviews of *The Age of Surveillance Capitalism*. *New Media & Society*, 23(9), pp. 2839–51.

Jargon, Julie (2021). Kids' chores starting to bore? New apps, assistants and smart appliances can motivate them. *Wall Street Journal*, 18 August, https://www.wsj.com/articles/kids-chores-starting-to-bore-new-apps-assistants-and-smart-appliances-can-motivate-them-11597752001.

Jethani, Suneel (2021). *The Politics and Possibilities of Self-Tracking Technology: Data, Bodies and Design*. Bingley: Emerald.

Jethani, Suneel and Dale Leorke (2021). *Openness in Practice: Understanding Attitudes to Open Government Data*. London: Springer Nature.

Jones, Rodney and Christof Hafner (2012). *Understanding*

Digital Literacies: A Practical Introduction. London: Routledge.

Jones, Sam (2020). Switzerland halts rollout of 5G over health concerns. *FT.com*, 12 February, https://www.ft.com/content/848c5b44-4d7a-11ea-95a0-43d18ec715f5.

Kantayya, Shalini (Director) (2020). *Coded Bias*. Sundance.

Kapadia, Anush (2020). All that is solid melts into code [book review]. *Economy and Society*, 49(2), pp. 329–44.

Kaye, D. Bondy Valdovinos, Xu Chen and Jing Zeng (2021). The co-evolution of two Chinese mobile short video apps: parallel platformization of Douyin and TikTok. *Mobile Media & Communication*, 9(2), pp. 229–53.

Kennedy, Helen (2016). *Post, Mine, Repeat: Social Media Data Mining Becomes Ordinary*. Basingstoke: Palgrave Macmillan.

Kennedy, Helen (2018). Living with data: aligning data studies and data activism through a focus on everyday experiences of datafication. *Krisis: Journal for Contemporary Philosophy*, 2018(1), pp. 18–30.

Kennedy, Helen and Giles Moss (2015). Known or knowing publics? Social media data mining and the question of public agency. *Big Data & Society*, 2(2), https://doi.org/10.1177/2053951715611145.

Kennedy, Helen and Martin Engebretsen (2020). Introduction: the relationships between graphs, charts, maps and meanings, feelings, engagements. In M. Engebretsen and H. Kennedy (eds.), *Data Visualisation in Society*. Amsterdam: Amsterdam University Press, pp. 19–32.

Kennedy, Helen and Rosemary L. Hill (2017). The pleasure and pain of visualizing data in times of data power. *Television & New Media*, 18(8), pp. 769–82.

Kennedy, Helen, Rosemary L. Hill, Giorgia Aiello and William Allen (2016). The work that visualisation conventions do. *Information, Communication & Society*, 19(6), pp. 715–35.

Kennedy, Jenny, Michael Arnold, Martin Gibbs, Bjorn Nansen and Rowan Wilken (2020). *Digital Domesticity: Media, Materiality, and Home Life*. New York: Oxford University Press.

Kennedy, Jenny and Rowan Wilken (2021). Liminoid media: on the enduring significance of USB portable flash drives. *New Media & Society*, 23(4), pp. 672–91.

Kidd, Dorothy (2019). Extra-activism: counter-mapping and data justice. *Information, Communication & Society*, 22(7), pp. 954–70.

Kinver, Mark (2020). Citizen science taps into public's love of trees. *BBC News*, 11 September, https://www.bbc.com/news/science-environment-54106793.

Kitchin, Rob (2021). *Data Lives: How Data are Made and Shape Our World*. Bristol: Bristol University Press.

Knobel, Michele and Colin Lankshear (eds.) (2008). *Digital Literacies Concepts, Policies and Practices*. New York: Peter Lang.

Koltay, Tibor (2011). The media and the literacies: media literacy, information literacy, digital literacy. *Media, Culture & Society*, 33(2), pp. 211–21.

Koopman, Colin (2019). *How We Became Our Data: A Genealogy of the Informational Person*. Chicago: Chicago University Press.

Krafft, Tobias D., Michael Gamer and Katharina Anna Zweig (2019). What did you see? A study to measure personalization in Google's search engine. *EPJ Data Science*, 8(38), https://doi.org/10.1140/epjds/s13688–019–0217–5.

Kukutai, Tahu and John Taylor (eds.) (2016). *Indigenous Data Sovereignty: Toward an Agenda*. Acton: ANU Press.

Langbauer, Laurie (1992). Cultural studies and the politics of the everyday. *Diacritics*, 22(1), pp. 47–65.

Learning and Work Institute (2021). Disconnected? Exploring the digital skills gap. Worldskills UK, London, https://learningandwork.org.uk/resources/research-and-reports/disconnected-exploring-the-digital-skills-gap.

Leaver, Tama (2017). Intimate surveillance: normalizing parental monitoring and mediation of infants online. *Social Media + Society*, 3(2), https://doi.org/10.1177/2056305117707192.

Leaver, Tama (2020). Sharenting, intimate surveillance, and the right to be forgotten. In L. Green, D. Holloway, K.

Stevenson, T. Leaver and L. Haddon (eds.), *The Routledge Companion to Digital Media and Children*. New York: Routledge, pp. 235–334.

Lefebvre, Henri (1987). The everyday and everydayness, trans. C. Levitch, A. Kaplan and K. Ross. *Yale French Studies*, 73, pp. 7–11.

Lefebvre, Henri (1991). *The Production of Space*, trans. D. Nicholson-Smith. Cambridge, MA: Blackwell.

Lefebvre, Henri (2000). *Critique of Everyday Life, Volume 1*, trans. J. Moore. London: Verso.

Lefebvre, Henri (2004). *Rhythmanalysis: Space, Time and Everyday Life*, trans. S. Elden and G. Moore. London: Continuum.

Light, Ben (2014). *Disconnecting with Social Networking Sites*. London: Palgrave Macmillan.

Light, Ben (2016). Producing sexual cultures and pseudonymous publics with digital networks. In R. A. Lind (ed.), *Race and Gender in Electronic Media: Content, Context, Culture*. London: Routledge, pp. 231–46.

Littler, Jo (2016). On not being at the CCCS. In K. Connell and M. Hilton (eds.), *Cultural Studies 50 Years On: History, Practice and Politics*. London: Rowman & Littlefield, pp. 275–84.

Livingstone, Sonia (2004). The challenge of changing audiences: or, what is the audience researcher to do in the age of the internet? *European Journal of Communication*, 19(1), pp. 75–86.

Livingstone, Sonia (2005). On the relation between audiences and publics. In S. Livingstone (ed.), *Audiences and Publics: When Cultural Engagement Matters for the Public Sphere*. Bristol: Intellect Books, pp. 17–41.

Livingstone, Sonia (2019). Audiences in an age of datafication: critical questions for media research. *Television & New Media*, 20(2), pp. 170–83.

Livingstone, Sonia and Peter Lunt (2016). Is 'mediatization' the new paradigm for our field? A commentary on Deacon and Stanyer (2014, 2015) and Hepp, Hjarvard and Lundby (2015). *Media, Culture & Society*, 38(3), pp. 462–70.

Lobato, Ramon (2019). *Netflix Nations: The Geography of Digital Distribution*. New York: New York University Press.

Lu, Jessica H. and Catherine Knight Steele (2019). 'Joy is resistance': cross-platform resilience and (re)invention of Black oral culture online. *Information, Communication & Society*, 22(6), pp. 823–37.

Lüders, Marika (2021). Pushing music: people's continued will to archive versus Spotify's will to make them explore. *European Journal of Cultural Studies*, 24(4), pp. 952–69.

Lupton, Deborah (2015). Quantified sex: a critical analysis of sexual and reproductive self-tracking using apps. *Culture, Health & Sexuality*, 17(4), pp. 440–53.

Lupton, Deborah (2016). *The Quantified Self*. Cambridge: Polity.

Lupton, Deborah (2018). Lively data, social fitness and biovalue: the intersections of health self-tracking and social media. In J. Burgess, A. Marwick and T. Poell (eds.), *The Sage Handbook of Social Media*. London: Sage, pp. 1–20.

Lupton, Deborah (2019). *Data Selves: More-than-Human Perspectives*. Cambridge: Polity.

Lupton, Deborah and Gavin J. Smith (2018). 'A much better person': the agential capacities of self-tracking practices. In B. Ajana (ed.), *Metric Culture: Ontologies of Self-tracking Practices*. Bingley: Emerald, pp. 57–75.

Lyons, Glenn and John Urry (2005). Travel time use in the information age. *Transportation Research Part A*, 39, pp. 257–76.

McCaskey, John P. (2020). History of 'temperature': maturation of a measurement concept. *Annals of Science*, 77(4), pp. 399–444.

McCosker, Anthony (2014). Trolling as provocation: YouTube's agonistic publics. *Convergence*, 20(2), pp. 201–17.

McCosker, Anthony (2017a). Data literacies for the post-demographic social self. *First Monday*, 22(10), http://firstmonday.org/ojs/index.php/fm/article/view/7307/6550.

McCosker, Anthony (2017b). Tagging depression: social media and the segmentation of mental health. In P. Messaris and L. Humphreys (eds.), *Digital Media: Transformations in Human Communication*. New York: Peter Lang, pp. 31–9.

McCosker, Anthony, Diana Bossio, Indigo Holcombe-James, Hilary Davis, Max Schleser and Jessamy Gleeson (2018). *60+ Online: Engaging Seniors through Social Media and Digital Stories*. Melbourne: Swinburne University of Technology.

McCosker, Anthony, Peter Kamstra, Tracy De Cotta, Jane Farmer, Frances Shaw, Zoe Teh and Arezou Soltani Panah (2021). Social media for social good? A thematic, spatial and visual analysis of humanitarian action on Instagram. *Information, Communication & Society*, 24(13), pp. 1870–90.

McCosker, Anthony and Rowan Wilken (2014). Rethinking 'big data' as visual knowledge: the sublime and the diagrammatic in data visualisation. *Visual Studies*, 29(2), pp. 155–64.

McCosker, Anthony and Rowan Wilken (2020). *Automating Vision: The Social Impact of the New Camera Consciousness*. New York: Routledge.

McCosker, Anthony and Timothy Graham (2018). Data publics: urban protest, analytics and the courts. *M/C Journal*, 21(3), https://doi.org/10.5204/mcj.1427.

McCosker, Anthony and Ysabel Gerrard (2021). Hashtagging depression on Instagram: towards a more inclusive mental health research methodology. *New Media & Society*, 23(7), pp. 1899–919.

Malpas, Jeff (1999). *Place and Experience: A Philosophical Topology*. Cambridge: Cambridge University Press.

Martin, Fran (2003). Introduction. In F. Martin (ed.), *Interpreting Everyday Culture*. London: Hodder Arnold, pp. 1–10.

Martins, Eduardo E. B. (2019). I'm the operator with my pocket vibrator: collective intimate relations on Chaturbate. *Social Media + Society*, 5(4), https://doi.org/10.1177/2056305119879989.

Marwick, Alice and danah boyd (2011). I tweet honestly, I tweet passionately: Twitter users, context collapse, and the imagined audience. *New Media & Society*, 13(1), pp. 114–33.

Marwick, Alice and danah boyd (2014). Networked privacy: how teenagers negotiate context in social media. *New Media & Society*, 16(7), pp. 1051–67.

Massey, Doreen (2005). *For Space*. London: Sage.

Masters, William and Virginia Johnson (1966). *Human Sexual Response*. First edition. Boston: Little, Brown & Company.

Meese, James, Jordan Frith and Rowan Wilken (2020). COVID-19, 5G conspiracies and infrastructural futures. *Media International Australia*, 177, pp. 30–46.

Mejias, Ulises A. and Nick Couldry (2019). Datafication. *Internet Policy Review*, 8(4), https://doi.org/10.14763/2019.4.1428.

Merrill, Jeremy and Will Oremus (2021). Five points for anger, one for a 'like': how Facebook's formula fostered rage and misinformation. *Washington Post*, 26 October, https://www.washingtonpost.com/technology/2021/10/26/facebook-angry-emoji-algorithm.

Metz, Rachel (2021). Likes, anger emojis and RSVPs: the math behind Facebook's news feed – and how it backfired. *CNN Business*, 27 October, https://edition.cnn.com/2021/10/27/tech/facebook-papers-meaningful-social-interaction-news-feed-math/index.html.

Micheli, Marina, Marisa Ponti, Max Craglia and Anna Berti Suman (2020). Emerging models of data governance in the age of datafication. *Big Data & Society*, 7(2), https://doi.org/10.1177/2053951720948087.

Miller, Daniel (2008). *The Comfort of Things*. Cambridge: Polity.

Miller, Daniel (2021). A theory of a theory of the smartphone. *International Journal of Cultural Studies*, 24(5), pp. 860–76.

Mitchell, Vanessa (2019). Zero party data: what on earth is it and why do marketers need it? *CMO*, 12 August, https://www.cmo.com.au/article/665141/zero-party-data-what-earth-it-why-do-marketers-need-it.

Møller Hartley, Jannie, Mette Bengtsson, Anna Schjøtt Hansen and Morten F. Sivertsen (2021). Researching publics in datafied societies: insights from four approaches to the concept of 'publics' and a (hybrid) research agenda. *New Media & Society*, https://doi.org/10.1177/14614448211021045.

Moran, Mayo (2003). *Rethinking the Reasonable Person: An Egalitarian Reconstruction of an Objective Standard*. Oxford: Oxford University Press.

Mosco, Vincent (2005). *The Digital Sublime: Myth, Power, and Cyberspace*. Cambridge, MA: MIT Press.

Mosseri, Adam (2021). How the 'algorithm' works. Instagram, 24 June, https://www.instagram.com/p/CQdxvdNJ_sC.

Mowlabocus, Sharif (2020). A Kindr Grindr: moderating race(ism) in techno-spaces of desire. In R. Ramos and S. Mowlabocus (eds.), *Queer Sites in Global Contexts*. London and New York: Routledge, pp. 33–47.

Moyo, Dumisani and Allen Munoriyarwa (2021). 'Data must fall': mobile data pricing, regulatory paralysis and citizen action in South Africa. *Information, Communication & Society*, 24(3), pp. 365–80.

Muir, Becca (2020). 'Armchair epidemiologists' and 'data bros': inside the DIY world of Covid-19 research. *Prospect*, 14 June, https://www.prospectmagazine.co.uk/science-and-technology/armchair-epidemiologists-and-data-bros-inside-the-diy-world-of-covid-19-research.

Murray, Brittany, Elsa Falkenburger and Priya Saxena (2015). *Data Walks: An Innovative Way to Share Data with Communities*. Washington, DC: Urban Institute, https://www.urban.org/research/publication/data-walks-innovative-way-share-data-communities.

Neff, Gina and Dawn Nafus (2016). *Self-Tracking*. Cambridge, MA: MIT Press.

Newman, Casey (2015). Twitter officially kills off favorites and replaces them with likes. *The Verge*, 3 November, https://www.theverge.com/2015/11/3/9661180/twitter-vine-favorite-fav-likes-hearts.

Newman, Christy et al. (2020). *Understanding Trust in Digital*

Health among Communities Affected by BBVs and STIs in Australia. Sydney: UNSW Centre for Social Research in Health, http://doi.org/10.26190/5f6d72f17d2b5.

Nieborg, David B. and Thomas Poell (2018). The platformization of cultural production: theorizing the contingent cultural commodity. *New Media & Society*, 20(11), pp. 4275–92.

Noble, Safiya Umoja (2018). *Algorithms of Oppression: How Search Engines Reinforce Racism.* New York: New York University Press.

North, Anna (2019). 7 positive changes that have come from the #MeToo movement. *Vox*, 4 October, https://www.vox.com/identities/2019/10/4/20852639/me-too-movement-sexual-harassment-law-2019.

Nye, David E. (1996). *American Technological Sublime.* Cambridge, MA: MIT Press.

Obungu, C. Brandon (2020). Don't be fooled by Covid-19 carpetbaggers. *Wired*, 4 May, https://www.wired.com/story/opinion-dont-be-fooled-by-covid-19-carpetbaggers.

O'Hara, Mary Emily (2019). Queer and feminist brands say they are being blocked from running ads on Instagram and Facebook, *MTV News*, 19 July, http://www.mtv.com/news/3131929/queer-and-feminist-brands-say-they-are-being-blocked-from-running-ads-on-instagram-and-facebook.

Öhman, Carl (2020). Introducing the pervert's dilemma: a contribution to the critique of Deepfake Pornography. *Ethics and Information Technology*, 22(2), pp. 133–40.

Ong, Walter J. (2013 [1982]). *Orality and Literacy.* New York: Routledge.

Ongweso Jr., Edward (2021). Amazon's new algorithm will set workers' schedules according to muscle use. *Vice Motherboard*, 16 April, https://www.vice.com/en/article/z3xeba/amazons-new-algorithm-will-set-workers-schedules-according-to-muscle-use.

Open Data Institute (2019). Data trusts: lessons from three pilots (report), 15 April, https://theodi.org/article/odi-data-trusts-report.

Paasonen, Susanna (2018). Affect, data, manipulation and price in social media. *Distinktion: Journal of Social Theory*, 19(2), pp. 214–29.

Pangrazio, Luciana (2016). Reconceptualising critical digital literacy. *Discourse: Studies in the Cultural Politics of Education*, 37(2), pp. 163–74.

Pangrazio, Luciana and Julian Sefton-Green (2020). The social utility of 'data literacy'. *Learning, Media and Technology*, 45(2), pp. 208–20.

Pangrazio, Luciana and Neil Selwyn (2018). 'It's not like it's life or death or whatever': young people's understandings of social media data. *Social Media + Society*, 4(3), https://doi.org/10.1080/2056305118787808.

Pangrazio, Luciana and Neil Selwyn (2019). 'Personal data literacies': a critical literacies approach to enhancing understandings of personal digital data. *New Media & Society*, 21(2), pp. 419–37.

Papacharissi, Zizi (2015). *Affective Publics: Sentiment, Technology, and Politics*. Oxford: Oxford University Press.

Park, Sora and Justine Humphry (2019). Exclusion by design: intersections of social, digital and data exclusion. *Information, Communication & Society*, 22(7), pp. 934–53.

Patton, Seth (2021). Microsoft Viva Insights helps people nurture wellbeing and be their best. *Microsoft Tech Community*, 2 April, https://techcommunity.microsoft.com/t5/microsoft-viva-blog/microsoft-viva-insights-helps-people-nurture-wellbeing-and-be/ba-p/2107010.

PenzeyMoog, Eva (2021). *Design for Safety*. A Book Apart.

Perec, Georges (1991). *Things: A Story of the Sixties* with *A Man Asleep*, trans. A. Leak. London: Harvill.

Perec, Georges (1999a). Approaches to what? In J. Sturrock (ed.), *Species of Spaces and Other Pieces*. Harmondsworth: Penguin, pp. 209–11.

Perec, Georges (1999b). Species of spaces. In J. Sturrock (ed.), *Species of Spaces and Other Pieces*. Harmondsworth: Penguin, pp. 1–96.

Perec, Georges (2010). *An Attempt at Exhausting a Place in Paris*, trans. M. Lowenthal. Cambridge: Wakefield Press.

Pertierra, Anna (2017). *Media Anthropology for the Digital Age*. Cambridge: Polity.

Petre, Caitlin, Brooke Erin Duffy and Emily Hund (2019). 'Gaming the system': platform paternalism and the politics of algorithmic visibility. *Social Media + Society*, 5(4), https://doi.org/10.1177%2F2056305119879995.

Pink, Sarah, Debora Lanzeni and Heather Horst (2018). Data anxieties: finding trust in everyday digital mess. *Big Data & Society*, 5(1), https://doi.org/10.1177/2053951718756685.

Pink, Sarah, Rosamaria Lucena, Jananda Pinto, Angélica P. C. De Souza, Camile Caminha, Geraldina Maria De Siqueira, Mariana Duarte de Oliveira, Alex Gomes and Renata Zilse (2019). Trust and knowing: emerging technologies and mobility in the Global South. In R. Wilken, G. Goggin and H. A. Horst (eds.), *Location Technologies in International Context*. New York: Routledge, pp. 173–87.

Pool, Ian (2016). Colonialism's and postcolonialism's fellow traveller: the collection, use and misuse of data on indigenous people. In T. Kukutai and J. Taylor (eds.), *Indigenous Data Sovereignty: Toward an Agenda*. Acton: ANU Press, pp. 57–78.

Prey, Robert (2018). Nothing personal: algorithmic individuation on music streaming platforms. *Media, Culture & Society*, 40(7), pp. 1086–100.

Prey, Robert (2020). Locating power in platformization: music streaming playlists and curatorial power. *Social Media + Society*, 6(2), pp. 1–11, https://doi.org/10.1177/2056305120933291.

Puschmann, Cornelius and Jean Burgess (2014). Metaphors of big data. *International Journal of Communication*, 8, pp. 1690–709.

Race, Kane (2015). 'Party and play': online hook-up devices and the emergence of PNP practices among gay men. *Sexualities*, 18(3), pp. 253–75.

Reichelt, Leisa (2007). Ambient intimacy. *Disambiguity* (blog), http://www.disambiguity.com/ambient-intimacy.

Reinstein, Julia (2018). 'Tweetdecking' is taking over Twitter. Here's everything you need to know. *BuzzFeed*, 12 January,

https://www.buzzfeednews.com/article/juliareinstein/
exclusive-networks-of-teens-are-making-thousands-of-
dollars.

Richardson, Ingrid, Larissa Hjorth, Yolande Strengers and William Balmford (2017). Careful surveillance at play: human-animal relations and mobile media in the home. In E. Gómez Cruz, S. Sumartojo and S. Pink (eds.), *Refiguring Techniques in Digital Visual Research*. London: Palgrave Macmillan, pp. 105–16.

Richman, Jessica (2019). This is the impact of Instagram's accidental fat-phobic algorithm. *Fast Company*, 15 October, https://www.fastcompany.com/90415917/this-is-the-impact-of-instagrams-accidental-fat-phobic-algorithm.

Robards, Brady, Ben Lyall and Claire Moran (2020). Confessional data selfies and intimate digital traces. *New Media & Society*, 23(9), pp. 2616–33.

Roberts, Sarah T. (2019). *Behind the Screen: Content Moderation in the Shadows of Social Media*. New Haven: Yale University Press.

Robertson, Craig (2021). *The Filing Cabinet: A Vertical History of Information*. Minneapolis: University of Minnesota Press.

Rosen, Jody (2014). The Knowledge, London's legendary taxi-driver test, puts up a fight in the age of GPS. *New York Times*, 10 November, https://www.nytimes.com/2014/11/10/t-magazine/london-taxi-test-knowledge.html.

Ruckenstein, Minna and Natasha Dow Schüll (2017). The datafication of health. *Annual Review of Anthropology*, 46, pp. 261–78.

Ruppert, Evelyn, Engin Isin and Didier Bigo (2017). Data politics. *Big Data & Society*, https://doi.org/10/1177/2053951717717749.

Sander, Ina (2020). Critical big data literacy tools – engaging citizens and promoting empowered internet usage. *Data & Policy*, 2, https://doi.org/10.1017/dap.2020.5.

Sandywell, Barry (2004). The myth of everyday life: toward a heterology of the ordinary. *Cultural Studies*, 18(2–3), pp. 160–80.

Schiffer, Zoe and Adi Robertson (2021). Watch a police officer admit to playing Taylor Swift to keep a video off YouTube. *The Verge*, 1 July, https://www.theverge.com/2021/7/1/22558292/police-officer-video-taylor-swift-youtube-copyright.

Schüll, Natasha (2021). Self-tracking. In N. Bonde Thylstrup, D. Agostinho, A. Ring, C. D'Ignazio and K. Veel (eds.), *Uncertain Archives: Critical Keywords for Big Data*. Cambridge, MA: MIT Press, pp. 457–68.

Schwartz, Raz (2015). Online place attachment: exploring technological ties to physical places. In A. de Souza e Silva and M. Sheller (eds.), *Mobility and Locative Media: Mobile Communication in Hybrid Spaces*. New York: Routledge, pp. 85–100.

Seaver, Nick (2017). Algorithms as culture: some tactics for the ethnography of algorithmic systems. *Big Data & Society*, 4(2), https://doi.org/10.1177%2F2053951717738104.

Seaver, Nick (2018). What should an anthropology of algorithms do? *Cultural Anthropology*, 33(3), https://journal.culanth.org/index.php/ca/article/view/ca33.3.04/89.

Sefton-Green, Julian (1998). *Digital Diversions*. London: UCL Press.

Shelby, Renee M. (2021). Documentary review: *Coded Bias*. *Society for Social Studies of Science*, https://www.4sonline.org/documentary-review-coded-bias.

Sheringham, Michael (2006). *Everyday Life: Theories and Practices from Surrealism to the Present*. New York: Oxford University Press.

Shome, Raka (2019). Thinking culture and cultural studies – from/of the Global South. *Communication and Critical/Cultural Studies*, 16(3), pp. 196–218.

Shun-Hing, Chan (2010). Interfacing feminism and cultural studies in Hong Kong: a case of everyday life politics. *Cultural Studies*, 16(5), pp. 704–34.

Siles, Ignacio, Andrés Segura-Castillo, Ricardo Solís and Mónica Sancho (2020). Folk theories of algorithmic recommendations on Spotify: enacting data assemblages in the global South. *Big Data & Society*, 7(1), https://doi.org/10/1177.2053951720923377.

Silverstone, Roger (2005). The sociology of mediation and communication. In C. J. Calhoun, C. Rojek and B. S. Turner (eds.), *The Sage Handbook of Sociology*. London: Sage, pp. 188–207.

Simonite, Tom (2021). What really happened when Google ousted Timnit Gebru. *Wired*, 8 June, https://www.wired.com/story/google-timnit-gebru-ai-what-really-happened.

Smith, Harrison (2019). Metrics, locations, and lift: mobile location analytics and the production of second-order geodemographics. *Information, Communication & Society*, 22(8), pp. 1044–61.

Sofia, Zoë (2000). Container technologies. *Hypatia*, 15(2), pp. 181–201.

Spanos, Brittany (2019). Giddy up! Here's what you need to know about the Yeehaw Agenda. *Rolling Stone*, 8 March, https://www.rollingstone.com/music/music-features/welcome-to-the-yee-yee-club-bitch-805169.

Spotify (2021). Celebrate your unique listening style with Spotify's Only You in-app experience. *For the Record*, 2 June, https://newsroom.spotify.com/2021–06–02/celebrate-your-unique-listening-style-with-spotifys-only-you-in-app-experience.

Stark, Luke and Anna Lauren Hoffman (2019). Data is the new what? Popular metaphors and professional ethics in emerging data culture. *Journal of Cultural Analytics*, https://doi.org/10.31235/osf.io/2xguw.

Storey, John (2014). *From Popular Culture to Everyday Life*. London: Routledge.

Street, Brian V. (1984). *Literacy in Theory and Practice*. Cambridge: Cambridge University Press.

Strengers, Yolande and Jenny Kennedy (2020). *The Smart Wife: Why Siri, Alexa, and Other Smart Home Devices Need a Feminist Reboot*. Cambridge, MA: MIT Press.

Sundén, Jenny (2020). Play, secrecy and consent: theorizing privacy breaches and sensitive data in the world of networked sex toys. *Sexualities*, https://doi.org/10.1177/1363460720957578.

Sung, Morgan (2020). A guide to using TikTok's algorithm to watch videos you actually like. *Mashable*, 14 January, https://mashable.com/article/tiktok-algorithm-watch-what-you-actually-like.

Swan, Melanie (2012). Sensor mania! The internet of things, wearable computing, objective metrics, and the quantified self 2.0. *Journal of Sensor and Actuator Networks*, 1(3), pp. 217–53.

Swart, Joëlle (2021). Experiencing algorithms: how young people understand, feel about, and engage with algorithmic news selection on social media. *Social Media + Society*, 7(2), https://doi.org/10.1177/20563051211008828.

Thomas, Julian (2018). Programming, filtering, adblocking: advertising and media automation. *Media International Australia*, 166(1), pp. 34–43.

Thomas, Julian et al. (2018). *Measuring Australia's Digital Divide: The Australian Digital Inclusion Index 2018*. Melbourne: RMIT University, for Telstra, https://digital inclusionindex.org.au/the-index-report/report.

Thorp, Jer (2021). *Living in Data: A Citizen's Guide to a Better Information Future*. New York: Farrar, Straus and Giroux.

Tiefer, Leonore (1991). Historical, scientific, clinical and feminist criticisms of 'the human sexual response cycle' model. *Annual Review of Sex Research*, 2(1), pp. 1–23.

Tiffany, Kaitlyn (2021). I'm scared of the person TikTok thinks I am. *The Atlantic*, 22 June, https://www.theatlantic.com/technology/archive/2021/06/your-tiktok-feed-embarrassing/619257.

Tiidenberg, Katrin, Natalie Ann Hendry and Crystal Abidin (2021). *Tumblr*. Cambridge: Polity.

Turner, Graeme (1990). *British Cultural Studies: An Introduction*. London: Routledge.

Van der Vlist, Fernando N. and Anne Helmond (2021). How partners mediate platform power: mapping business and data partnerships in the social media ecosystem.

Big Data & Society, 8(1), https://doi.org/10.1177/20539517211025061.

Van Dijck, José and Thomas Poell (2013). Understanding social media logic. *Media and Communication*, 1(1), pp. 2–14.

Van Dijck, José, Thomas Poell and Martijn de Waal (2018). *The Platform Society: Public Values in a Connective World*. Oxford: Oxford University Press.

Varghese, Sarah (2019). Pokémon Go was a warning about the rise of surveillance capitalism. *Wired*, 3 February, https://www.wired.co.uk/article/the-age-of-surveillance-capitalism-facebook-shoshana-zuboff.

Varon, Joanna (2020). The Future Is TransFeminist: from imagination to action. *DeepDives*, https://deepdives.in/the-future-is-transfeminist-from-imagination-to-action-6365e097eb22.

Vee, Annette (2017). *Coding Literacy: How Computer Programming is Changing Writing*. Cambridge, MA: MIT Press.

Vincent, James (2021a). Tom Cruise deepfake creator says public shouldn't be worried about 'one-click fakes'. *The Verge*, 5 March, https://www.theverge.com/2021/3/5/22314980/tom-cruise-deepfake-tiktok-videos-ai-impersonator-chris-ume-miles-fisher.

Vincent, James (2021b). Amazon delivery drivers have to consent to AI surveillance in their vans or lose their jobs. *The Verge*, 24 March, https://www.theverge.com/2021/3/24/22347945/amazon-delivery-drivers-ai-surveillance-cameras-vans-consent-form.

Vincent, James (2021c). Amazon denies stories of workers peeing in bottles, receives a flood of evidence in return. *The Verge*, 25 March, https://www.theverge.com/2021/3/25/22350337/amazon-peeing-in-bottles-workers-exploitation-twitter-response-evidence.

Vinsel, Lee (2021). You're doing it wrong: notes on criticism and technology hype. *Medium*, 2 February, https://sts-news.medium.com/youre-doing-it-wrong-notes-on-criticism-and-technology-hype-18b08b4307e5.

Vivienne, Son (2016). *Digital Identity and Everyday Activism: Sharing Private Stories with Networked Publics*. London: Palgrave Macmillan.

Vygotsky, Lev S. (1980). *Mind in Society: The Development of Higher Psychological Processes*. Cambridge, MA: Harvard University Press.

Watkins, Evan (2015). *Literacy Work in the Reign of Human Capital*. New York: Fordham University Press.

Weed, Julie (2019). 'These Apps Are an Uber Driver's Co-Pilot'. *New York Times*, 17 October, https://www.nytimes.com/2019/10/17/business/apps-uber-lyft-drivers.html.

Weiss, Haley (2018). Why people love Spotify's annual wrap-ups. *The Atlantic*, 13 December, https://www.theatlantic.com/technology/archive/2018/12/spotify-wrapped-and-data-collection/577930.

Wikström, Patrik (2019). *The Music Industry: Music in the Cloud*. Third edition. Cambridge: Polity.

Wilken, Rowan (2017). The quick brown fox jumps over the lazy dog: Perec, description and the scene of everyday computer use. In R. Wilken and J. Clemens (eds.), *The Afterlives of Georges Perec*. Edinburgh: Edinburgh University Press, pp. 226–42.

Wilken, Rowan (2019). *Cultural Economies of Locative Media*. New York: Oxford University Press.

Wilken, Rowan, Jean Burgess and Kath Albury (2019). Dating apps and data markets: a political economy of communication approach. *Computational Culture*, 7, http://computationalculture.net/dating-apps-and-data-markets-a-political-economy-of-communication-approach.

Wilken, Rowan and Jenny Kennedy (2022). Everyday data cultures and USB portable flash drives. *International Journal of Cultural Studies*, 25(2), pp. 192–209.

Williams, Linda (1989). *Hard Core: Power, Pleasure and the 'Frenzy of the Visible'*. Berkeley and Los Angeles: University of California Press.

Williams, Raymond (1961 [1958]). *Culture and Society 1780–1950*. Harmondsworth: Pelican.

Williams, Raymond (1975 [1961]). *The Long Revolution*. Harmondsworth: Penguin.

Williams, Raymond (1976). *Keywords: A Vocabulary of Culture and Society*. London: Croom Helm.

Williams, Raymond (1989). *Resources of Hope: Culture, Democracy, Socialism*. London: Verso.

Williams, Sarah (2020). *Data Action: Using Data for Public Good*. Cambridge, MA: MIT Press.

Winter, Rachel and Anastasia Salter (2020). DeepFakes: uncovering hardcore open source on GitHub. *Porn Studies*, 7(4), pp. 382–97.

Wolfe, Natalie (2018). A new app is about to change the way we have sex – and it all has to do with the #MeToo movement. *News.com.au*, 12 January, https://www.news.com.au/technology/online/security/a-new-app-is-about-to-change-the-way-we-have-sex-and-it-all-has-to-do-with-the-metoo-movement.

Wolmark, Jenny (ed.) (1999). *Cybersexualities: A Reader on Feminist Theory, Cyborgs and Cyberspace*. Edinburgh: Edinburgh University Press.

Wood, Robert W. (2020). Uber & Lyft ordered to treat drivers as employees, are any contractors independent now? *Forbes*, 11 August, https://www.forbes.com/sites/robertwood/2020/08/11/uber--lyft-ordered-to-treat-drivers-as-employees-are-any-contractors-independent-now/?sh=6f5f50285516.

Woodcock, Jamie (2020). The algorithmic panopticon at Deliveroo: measurement, precarity, and the illusion of control. *Ephemera*, 20(3), pp. 67–95.

Woodcock, Jamie and Mark Graham (2019). *The Gig Economy: A Critical Introduction*. Cambridge: Polity.

Yao, Xiaofang, Anthony McCosker, Kath Albury, Alexia Maddox and Jane Farmer (2021). Building data capacity in the not-for-profit sector: interim report. Melbourne: Swinburne University of Technology, https://apo.org.au/node/314477.

YouTube (2021). How does YouTube prevent copyright piracy on the platform?, https://www.youtube.com/

howyoutubeworks/our-commitments/safeguarding-copyright.

Ytre-Arne, Brita and Hallvard Moe (2021). Folk theories of algorithms: understanding digital irritation. *Media, Culture & Society*, 43(5), pp. 807–24.

Yuan, Xiaoyi, Ross J. Schuchard and Andrew T. Crooks (2019). Examining emergent communities and social bots within the polarized online vaccination debate in Twitter. *Social Media + Society*, 5(3), https://doi.org/10.1177%2F2056305119865465.

Zuboff, Shoshana (2019). *The Age of Surveillance Capitalism: The Fight for a Human Future at the New Frontier of Power*. London: Profile Books.

Zwitter, Andrej (2014). Big data ethics. *Big Data & Society*, 1(2), https://doi.org/10.1177/2053951714559253.

Index